I0479846

Mushroom Cultivation

Instruction for growing mushroom at home with safe uses

Isabella Taylor

TABLE OF CONTENT

ABOUT THE AUTHOR

Isabella Taylor now a woman, when she was a child she went to the english countryside with her grandfather to the harvest of mushrooms and truffles, this passion remained even after the disappearance of her grandfather, she graduated in biology and chemistry. still continues his passion and teaches the paseo perosne how to grow their mushrooms. start this journey with this guide and let yourself be carried away by isabella's passion.

© Copyright 2019 by Isabella Taylor all rights reserved. This document is geared towards providing exact and reliable information with regards to the topic and issue covered. The publication is sold with the idea that the publisher is not required to render accounting, officially permitted, or otherwise, qualified services. If advice is necessary, legal or professional, a practiced individual in the profession should be ordered. - From a Declaration of Principles which was accepted and approved equally by a Committee of the American Bar Association and a Committee of Publishers and Associations. In no way is it legal to reproduce, duplicate, or transmit any part of this document in either electronic means or in printed format. Recording of this publication is strictly prohibited and any storage of this document is not allowed unless with written permission from the publisher. All rights reserved. The information provided herein is stated to be truthful and consistent, in that any liability, in terms of inattention or otherwise, by any usage or abuse of any policies, processes, or directions contained within is the solitary and utter responsibility of the recipient reader. Under no circumstances will any legal responsibility or blame be held against the publisher for any reparation, damages, or monetary loss due to the information herein, either directly or indirectly. Respective authors own all copyrights not held by the publisher. The information herein is offered for informational purposes solely, and is universal as so. The presentation of the information is without contract or any type of guarantee assurance. The trademarks that are used are without any consent, and the publication of the trademark is without permission or backing by the trademark owner. All trademarks and brands within this book are for clarifying purposes only and are the owned by the owners themselves, not affiliated with this document

INTRODUCTION

The mushrooms represent a group of living organisms comparable to very atypical plants: in fact, unlike the latter, they are devoid of chlorophyll and differ from most plants because they need to live on substances already processed by other living beings, as they are not able to process them or make them yourself. They resemble green plants because, with a few exceptions, they have defined cell walls and, just like plants, are immobile. Finally they reproduce by means of spores, which can be compared to the seeds of higher plants. However, the mushrooms have no stem, roots or leaves and are devoid of the vascular system, which from the roots carries the lifeblood to go up the trunk until it reaches the branches and leaves, typical of plants. Like any living organism, they too are formed by the set of an indefinite number of cells, where by cell we mean the basic system of the structure and functioning of every living organism. The main part of the mushroom is formed by a thin and intricate network of whitish filaments, most of the time invisible to the naked eye, which, starting from the base of the stem, branch off into the soil and soil below, sometimes even for several tens of meters of length.

The mushrooms are part of the plant kingdom of our earth - so it was thought earlier. Today we know that mushrooms are neither plants nor animals. In biology, mushrooms are the third largest realm of eukaryotic organisms. Like animals (humans also belong to them), fungi are heterotrophic, so they need organic nutrients formed by other living things for their metabolism, which they usually unlock through the release of enzymes and thus make them available to them. Fungi, like plants, are not capable of photosynthesis based on chlorophyll (leaf green) and thus extract their life energy from sunlight. This brings them closer to the animals.

Another common feature of fungi and animals is that both form the polysaccharide glycogen as a storage substance, while plants form starch. The polysaccharide cellulose is characteristic of plants, which is absent from fungi. The cell walls of fungi mainly consist of chitin, which moves them closer to the animals. The whole metabolism and genetic factors also move the mushrooms closer to the animals.

The immobility of the mushrooms is not a criterion to differentiate them from the animals. The sponges or stone corals spend most of their lives stationary and are still animals. There are significant differences between fungi and animals in the ultrastructure, such as the presence of solid cell walls and vacuoles (expandable cell organelles) - animal cells do not have that. In addition, mushrooms reproduce very differently than animals.

These organisms are versatile. They occur as boletus in the forest, as truffles underground, but also as a skin fungus on our body surface. There they invisibly populate us. Under the electron microscope, for example, some molds look something like a dandelion.

In addition to animals and plants, the fungi are the third largest organism among eukaryotes.

Most people think of them as plants because they are typically earth-bound like plants and not like animals moving and looking for food; In the grocery stores you can find the edible mushrooms such as mushrooms and shiitake in the vegetable section. However, no fungus is capable of photosynthesis itself, its energy metabolism is always driven by organic molecules from dead and sometimes living organisms. In fact, according to DNA comparisons, fungi are more closely related to animals than to plants. After that, the line of development that led to us (and all other animals) separated later from the one that led to the truffles (and all other mushrooms) than that that led to the green plants. However, as estimates of mutation rates (the "molecular clock") suggest, this happened perhaps 800 million years ago - deep in the Precambrian. The earliest fossil finds of mushrooms are much later, around 400 million years ago, together with the fossils of the first land plants.

There are also mushrooms in the sea and in fresh water, but most live in the countryside, where they are literally ubiquitous and have developed an unmanageable variety of food sources. Their number of species is estimated at perhaps 1.5 million, but less than a tenth of them have been scientifically described to date.

An outward digestive organ

Fungi break down their food through excreted enzymes outside their body and absorb the nutrients in dissolved form through their cell membranes. This corresponds to the type of food we eat in our small intestine. The largest possible absorption surface is crucial for effective digestion and absorption of food. In the intestine, this is achieved through the villi (villi) and the brush border of the microvilli of the epithelial cells. A common field mushroom achieves a comparable absorption area through a kilometer-long network of threads (hyphae) with which it runs through the soil. The entire hyphal network of a fungus is called mycelium. What is commonly perceived as a fungus is only a small part of the organism, namely the fruiting body in which the spores are formed. Unlike animals and plants, mushrooms do not form embryos. Rather, they grow out of single-celled, tiny spores that can be formed sexually or asexually, often in enormous quantities. With every breath we take in fungal spores that are spread through the air.

Most people are aware that mushrooms only play a subordinate role compared to plants and animals. In fact, thanks to their ability to convert organic residues into mineral resources, they play a crucial role in the global material cycles. In terms of mass, probably around a quarter of the total biomass on our planet, they far exceed the total mass of all animals. In a pine forest in Michigan, USA, for example, a Hallimasch (Armillaria bulbosa) was found on the tree roots, the mycelium that apparently emerged from a single spore spreads over 37 hectares and weighs an estimated 11 tons. That it is actually a single organism

False and real mushrooms

Not all fungi form mycelia. Some, like yeast, are single-celled organisms that divide as a whole or by budding and pinching off spores and growing as a diffuse mass. The transition from unicellularity to multicellularity (and probably also the reverse process) has occurred several times in the realm of mushrooms.

In the past, many different mushroom-growing and nourishing, chlorophyll-free organisms were classified as fungi. In the meantime, molecular markers and, above all, genetic analyzes have led to a system that is more likely to do justice to the natural relationships. There is no doubt that the so-called radiation fungi (Actinomycetes), although they form mycelia from cell filaments and actin spores, have nothing to do with the fungi. They are prokaryotes (henceforth renamed Actinobacteria). Its best-known representative, Streptomyces, is, however, the most important producer of antibiotics alongside the penicillium and relatives, which are genuine fungi. What used to be classified as slime molds (including Dictyostelium and Physarum.

The position of the Oomycetes, which includes important plant pests such as Phytophthora infestans, which causes potato blight, or Plasmopara viticola, the downy mildew of the grapevine, is controversial. In contrast to the real mushrooms, their cell wall contains cellulose like the green plants and they form spores that can move in the water with a scourge. The little-known chytridia, which form flagellated reproductive cells (asexual zoospores and sexually produced gametes) and sometimes contain cellulose in addition to chitin, occupy an intermediate position. Real mushrooms, on the other hand, do not have cellulose, but chitin as the main component of their cell wall. They never have flagellated stages and their sexual reproduction does not take place through the fusion of single cells (gametes), but through conjugation of different hyphae, which, in the absence of morphological distinguishing features, are not called female and male, but rather plus and minus. This sexual act gives rise to the fungal fruiting bodies in which spores are formed by meiosis.

The real mushrooms include the zygomcetes or yoke mushrooms, including the common bread mold (Rhizopus stolonifer). The cotton rot on strawberries is also caused by this fungus. The huge group of "higher mushrooms" is divided into two large groups, which differ mainly in their fruiting bodies, in which the spores are formed:

I. *The Ascomycetes or tubular mushrooms*: In addition to coveted edible mushrooms such as truffle and morel, they include numerous other species that are useful and harmful to humans, including the ergot claviceps (alkaloids), the molds Aspergillus (aflatoxins, statins), penicillium (antibiotics) and tolypocladium (cyclosporin) as well as the beer, Wine and baker's yeast Saccharomyces). There are many specialists among the Ascomycetes who can break down the most resistant substances of the animal and plant kingdom, such as cellulose and the lignin of wood, the keratin of the hair and nails and the collagen of the bones and connective tissue. They are also of great importance for the industrial biotechnological production of enzymes and organic compounds such as citric acid or amino acids,

II. *The basidiomycetes, the stand or hat mushrooms:* They include most of the known edible and poisonous mushrooms, including the hallucinogenic mushrooms such as Psilocybe or Amanita, which are sought after or feared depending on their point of view. Some serve as a source of potent active ingredients for the pharmaceutical and agro industries. The class of strobilurins used in crop protection was first discovered in small turnips growing on pine cones. Dreaded plant pests such as rust and smut fungi also belong to the Basidiomycetes. Numerous species form stable symbioses with the roots of woody plants (mycorrhiza), without which an adequate supply of nutrients to the plants from the ground would not be possible.

When discussing the mushrooms, the lichens, hermaphrodites from fungi and algae, must at least be mentioned. They represent perhaps the most highly developed form of a symbiosis and are ecologically particularly important as first settlers in the most inhospitable places on earth.

Sexual reproduction does not take place in many mushrooms (or at least has not been observed). Since they cannot be classified as Ascomycetes or Basidiomycetes due to the lack of fruiting bodies, they are summarized as Deuteromycetes or Fungi imperfecti (imperfect fungi).

Medical meaning of mushrooms

Among the approximately 150 fungi known to be pathogenic to humans, many belong to these fungi imperfecti, including Candida albicans, which primarily affects the genital area, and the causative agents of the skin, hair and nail mycoses, which are summarized as dermatophytes. While these human pathogenic fungi are generally to be regarded as unpleasant rather than dangerous, they can pose a great danger to the defenseless individual.

But not only fungal infections, but already dead fungal cells or their components can trigger allergies as foreign antigens and pose a health hazard. Fungal antigens are the most common allergens for humans. Poisoning from fungi does not only exist in the acute form as after the consumption of tuberous mushrooms. Around 2,500 mycotoxin-producing fungi are known, far more than infectious species. Mycotoxins with organotoxic, mutagenic and teratogenic properties are known, and their importance for the development of cirrhosis of the liver has not yet been researched. There is still a wide field for science here.

MUSHROOM PROPERTIES

Fungi contribute to the proper functioning of our immune system. Their calorie intake is not high. For all varieties of mushrooms it is around 25 kcal per 100 g of fresh product, which can contain water up to 90% of its weight . In a 100 gr portion of fresh mushrooms there are 4.5 gr of carbohydrates, 3.5 gr of protein and 0.3 gr of fat.

Mushrooms are considered a good source of mineral salts such as potassium, phosphorus, copper and selenium, all essential for the proper functioning of our body. The ancients must have already been aware of the properties of mushrooms, so much so that in Egypt they were considered a particularly delicious and beneficial food, worthy of a pharaoh.

Mushrooms contain lysine and tryptophan , as far as proteins are concerned. In them there is also the presence of B vitamins and antioxidant substances, considered useful for the prevention of tumor pathologies and diseases related to aging. Mushrooms are useful in fighting cardiovascular diseases and to stem cholesterol build-up in the arteries.

For some, mushrooms are mistakenly thought of as a nutrient-poor food. This has been denied by the finding within them of selenium, whose nutritional contribution allows us to defend ourselves both from infectious diseases and from the appearance of tumors. They also contain vitamin B3, which contributes to the proper functioning of the nervous system and proper oxygenation of the blood, and vitamin B2, necessary both for the production of red blood cells and for the metabolism of proteins, fats and carbohydrates.

As for their power to strengthen the immune system, mushrooms have been considered for centuries as a real natural antibiotic and are indicated by unconventional medicine as a precious food to be taken during the change of season to protect themselves from autumnal ailments. Their appearance in the woods between the end of summer and the beginning of autumn, or between winter and spring, confirms that it is precisely the passage between the seasons that is the ideal period to consume them fresh and to enjoy all the benefits they present.

THR ORIGIN OF MUSHROOM

It may seem strange and out of place to write about mushrooms in a marine biology site, in fact, mushrooms are often associated with different and particular environments and ecosystems, for example mountain and wooded areas, environments therefore far away, as a typology, from marine ones in general ; but not everyone knows that they too have a curious and particular evolutionary history and, as we will see later, linked precisely to the aquatic environment.

we have dealt with peat bog ecosystems, noting in particular the importance they have as bioindicators of plant species and consequently climates that have alternated in the past. Among the various inclusions that are hidden between their sediments, in addition to spores, pollen and anything else, there are also testimonies of mushrooms, certainly not the mushrooms as we understand them, in fact they are microscopic mushrooms and linked through mycorrhiza symbiosis to different types of aquatic plants, most of the times now extinct. Mycorrhizae are symbiotic associations that are established between a plant and several mushroom taxa.

And it is precisely from the fossil bogs that their long evolutionary history and their genesis begins mycorrhizae

in this regard, that the mycorrhizae are divided into ectotrophic (image on the side) and endotrophic : the former develop on roots and trees, forming a network of filaments and, some, showy fruiting bodies (carpophores), also hypogean (for example truffles); the filaments of the ectotrophic mycorrhizae never penetrate the plant cells. Endotrophic mycorrhizae develop within the roots and are not visible to the naked eye. In this case the hyphae penetrate the plant cells without however damaging them. They are widespread in cold and temperate woods and transfer the decomposition products of the leaf litter to the plants with which they enter into symbiosis.

Although the most ancient fossil remains can be difficult to interpret and catalog, and this is also due to the fact that the mushrooms have a soft body and therefore fossilize with extreme difficulty, scholars agree that this group of organisms is very ancient , although they can be based solely on the examination of a few elementary structures.

The discovery discovered at the beginning of the last century in a peat deposit located near Rhynie, a Scottish locality whose sediments are dated 350 million years ago, is very interesting. It is a swamp of the Middle Devonian period, in which a fossil flora of particular interest has been perfectly preserved. The marsh plants unearthed in Rhynie are almost all very different from those known at the time, although they also had similar conducting tissues, produced spores and possessed tuberous structures at the level of the roots. But the most interesting fact is that when scholars examined their rhizomes under a microscope, they made a further amazing discovery: they contained fungal mycelia. Rhynie fossils have survived to us thanks to silicization, process where the organic components have been replaced by silica from residual thermal sources of volcanic activity, located nearby. The conservation of these fossil finds therefore took place in flint veins (calledchert), within the surrounding sediments.

Some of the plants present in this field have fungal traces. These include Aglaophyton , a very primitive pseudovascular plant with conduction cells similar to certain mosses. It had no leaves or roots. Or Horneophyton , a vascular plant with branched, bifurcated sporangia and vascular tissue only partially distributed along the stems. In both species microscopic mushrooms belonging to the group of chitridae have been found (from the Greek chytridion , which means small pot)micro fungi.

Aquatic species, flagellated algae parasites and various types of plants belong to this division of mushrooms. The chitrids have cenocytic hyphae (cellular organization with multiple nuclei, which derive either from the division of an initial nucleus or from the fusion of multiple cells) or poorly septated and uniflagellate spores (zoospores). They are primitive saprophytes, in fact they degrade chitin and keratin.

The evolution of mushrooms continues also in the following period of the Paleozoic era, in fact in the Carboniferous, about 300 million years ago, there are findings of archaic basidiomycete mushrooms, ascribed to the genus Archagaricon and coming from Northumberland (England), headquarters, between the other, of important coal deposits; moreover, various pedunculate forms, forming shelves with a honeycomb-shaped fruiting body and pores ascribed to the Polyporus genus , are found in the tertiary soils of the Libyan desert. Italy too has a similar testimony, more precisely in Veneto, in the famous reservoir of the Chiavon (VI) stream, belonging to the middle Oligocene period (about 30 million years ago); the first known specimen of was foundFossil Agaricus . It is certainly not to be excluded, moreover, that fungal specimens may be found in peat deposits in the near future, in the various paleolags existing in some fossil locations of various periods, belonging to the Cenozoic and Neozoic eras.

Curiosity and scientific interest, has recently aroused the discovery of rare mushroom species incorporated in fossil resins and amber. The term amber is used to identify the fossilized resin of trees that are now extinct which, dripping from the stems of the latter, incorporated everything it encountered in its descent downwards, including insects, mosses, leaves and flowers, but also rare mushrooms. Subsequently, the resin solidified through various chemical and physical steps and processes which, on the whole, are called amberization, preserving up to us as in a casket, all the organic remains of the past. In this regard, lately there is also an amber sample on the internet, including a basidiomycete mushroom, whose monetary evaluation, among other things, is very high Amber with mushroom

A curiosity: some mushrooms belonging to the Chytriodiomycota division (flagellated mushrooms) have been described thanks to the observation of some fossil teeth belonging to rhinos, found in German Miocene swamps, which had passages and furrows dug precisely by filamentous mushrooms which, after the death of the animal, they settled on their tooth substance.

we can say that mushrooms are eukaryotic organisms, such as algae and as they show a large amount of forms, from single-celled to filamentous, as well as large forms and, like algae, mushrooms could have characters in common, al point that some authors propose a hitherto unknown common progenitor from which, later, they would both have taken different evolutionary paths.

GENERAL CHARACTERISTICS OF MUSHROOM

About 75,000 species have been recognized but they are thought to be in fact between 800,000 and 1,500,000. Let's analyze the general characteristics by focusing on thallus, specialized vegetative structures and reproductive structures.

The thallus

It represents the vegetative body, this is generally filamentous and composed of filaments called hyphae, which make up the mycelium. We also distinguish yeast- like thalli. There are also mushrooms that instead can change from one type to another when environmental conditions vary, they are called dimorphic mushrooms.

The walls that delimit the thallus of a mushroom are very thin and have complex layered and chemical structures, these are equipped with fibrillary polymers inside and amorphous polymers to form the matrix. The main component is chitin, an N-acetyl glucosamine polymer, organized in microfibers.

Other components are alpha-glucans and chitosans. In yeasts chitin is scarcely present, therefore mannans prevail . It is also absent in oomycetes , where a cellulose-like polymer prevails organized not in fibers but in an amorphous matrix, as in algae (oomycetes are the "fungi" similar to heteroconte algae).

The cells constituting the hyphae can be multinucleated, said asettate or cenocitiche , or septate . In ascomycetes and in most anamorphic mushrooms the septum is not complete but has a central pore which allows cytoplasmic continuity; there are the bodies of Woronin that occlude the pores in case of damage to the hyphae, to isolate each cell. In basidiomycetes the septum is called dolipore , whose central pore is small enough to prevent the migration of organelles, the ends of the septum are swollen and surmounted by membranous hemispherical hoods, called parenthesomes .

Peculiarities of mushrooms

Ergosterol, this replaces the role and functions that cholesterol performs in animals

- Mushroom nuclei are much smaller than plant and animal nuclei
- chromosomes often do not condense during mitosis
- centrioles are present only in groups with flagellated elements
- the tubulins are different from the animal and vegetable ones
- Specialized vegetative structures and reproductive structures

Mushrooms only apical growth, new walls are synthesized only at the apex, where there are vesicles of the wall that blend with the apical plasma membrane pouring its contents on the outside of the rising wall. Below the apex, the hyphae tend to branch going to occupy the free spaces forming colonies. Several hyphae can organize together to form more or less complex structures, such as mycelial cords and sclerotia , reserve and resistance structures that develop with the compaction of hyphae and sometimes fusion (anastomosis).

In mature sclerotia, the external hyphae are swollen and curly with melanin. Portions of fragmented mycelium can generate new mycelia under favorable conditions. Mushrooms, however, can reproduce both asexually and sexually through the production of spores. In the phyla Oomycota , Chytridiomycota and Zygomycota the asexual spores are sporangiospores , that is, produced within a closed apical structure, the sporangium . The sporangium is delimited by the underlying hypha by a septum, initially it is multinucleated but, differentiating itself through a process called cleavage, membranes are formed, and the nuclei become spores.

In the zoosporic fungi (Oomycota , Chitridiomycota , Plasmodiophoromycota , the sporangiospores are mobile, equipped with flagella and without walls (zoospores); in the Zygomycota instead, they have no flagella, they are immobile and equipped with a wall. In the phyla Ascomycota and Basidiomycota and in the anamorphic mushrooms the asexual spores are conidia , spores not produced, that is, within a conidiangio , not flagellated and with walls.

A modality of conidiogenesis sees an hyphal tract settling intensely up to fragmenting into coninical uninuclear segments. Another modality involves the vial cells (phialides) in a sort of budding: a primordium is produced as swelling at the apex of a phialidium which then matures in conidium when delimited by a septum. The process is repeated and the conidia adhere in loose chains carried on erect structures (conidiophores). This helps the air dispersion which is typical for mushrooms.

Sexual spores, on the other hand, are produced differently in each phylum. The most complex way is found in large sporophores of ascomycetes and basidiomycetes, which can have hyphae specializing in generative , skeletal and connective tissue.

Genetic variability mechanisms

Generally the nuclei of the fungi are haploid, a condition which can however be complemented by sequences present in other nuclei; in fact the real mushrooms are considered functional diploids. The co-presence of genetically different nuclei is defined as heterocaryosis , while a colony containing only genetically equal nuclei is called homocaryotic .

The homocaryotic situation can be restored through the production of uninucleate spores or with the formation of a branch through which only one type of nucleus passes which carries out mitosis. Many mushrooms have sexual reproduction, but discrete single-cell gametes are almost always not produced, rather two hyphae can merge, provided they are compatible. The hyphae are not distinct in male or female but can be genetically different, i.e. of different mating-type (type of sexual polarity) indicated as + and - , only two opposite signs can merge.

However, plasmogamy is not immediately followed by cariogam , in fact in some fungi it is delayed and hyphae containing pairs of the two nuclear types (dicarion) can form which will proliferate by synchronous mitosis. These hyphae may merge with other dicaryotic hyphae but in the end each pair will merge and give a diploid zygotic nucleus, which will undergo meiosis. In other fungi (Aspergillus nidulans) there is a parasexual cycle , that is, two nuclei in a mycelium can merge to form a diploid nucleus that will have mitosis and, rarely, mitotic crossing-over. In subsequent mitoses, non-disjunction of chromosomes can occur, with the formation of nucleianeuploids that will lose other chromosomes in subsequent mitosis until the haploid condition is restored. If the parasexual cycle occurs on the nuclei that will become part of the conidia, these nuclei will have a different genetic makeup from the original nuclei and this could be the only mechanism of genetic recombination in the anamorphic mushrooms.

NUTRITION

All mushrooms are heterotrophic, they can take small molecules in solution by absorbing them from the wall, but they can also externally degrade more complex molecules through the secretion of digestive enzymes. They can also degrade substances of artificial origin (xenobiotics), therefore they are the largest known degraders. Depending on the state of the carbon they absorb, the fungi are divided into:

Saprotofi, feed on dead organic material, decompose cellulose and mineralize lignin; symbionts, deriving nourishment from living organisms, establishing mutually beneficial interactions with them, can be defined as mutualists or pathotists ;

Parasites, derive nourishment from other living beings with one-sided advantage and can be further distinguished in biotrophies (if they come into contact with the host cells without killing them) and in necrotrophies (if they kill and invade the host cells using the resources thus freed)

In addition to carbon, however, mushrooms also need other nutrients (macronutritive), with the exception of calcium (which is, only in mushrooms, micronutritive). Almost all mushrooms assimilate nitrogen in ammonia form, others can take and reduce nitrates and nitrites, while still others cannot use it in inorganic form and require amino acids.

Metabolism and physiology

Mushrooms have evolved in a very versatile way and are able to adapt to many different environmental conditions. Some can live both in the presence and absence of oxygen, thanks to the transition to a fermentative metabolism . Furthermore, mushrooms can implement a secondary metabolism suitable for the production of products not essential for growth but which can act as protection systems against external agents.

Often, the production of secondary metabolites has positive results for humans, since they can have great pharmacological capacities (such as penicillins , produced by Penicillium chrysogenum), but also negative (such as the deadly Amanita phalloides). The so-called mycotoxins , produced by some fungi, are very dangerous for humans. An example are aflatoxins , produced by Aspergillus flavus, which are the most carcinogenic substances known.

Being also deteriorating agents, they are also dangerous for their ability to tolerate extreme environmental conditions. Psychrophilic mushrooms grow at temperatures equal to or lower than 0 ° C and not more than about 20 ° C, while thermophilic mushrooms survive at temperatures equal to or greater than 50 ° C, with a lower limit of about 20 ° C and never above 62-65 ° C. Fungi are also the most xerophilous organisms that exist, they can survive in conditions of extreme dehydration. These mushrooms grow in dry substrates such as jellies, jams, grains, dried fruit, salted meats and fish, and produce large quantities of polyols in the cytoplasm (mannitol) whose hydroxyl groups replace part of the necessary hydration water.

Mushrooms similar to protozoa

Although studied by mycologists, their only affinity with them is the production of spores with walls, even if with a different chemical composition. In Myxomycota the thallus is formed by a multi-nucleated (plasmodium) wallless protoplasm that can migrate on the substrate like an amoeba. Plasmodium, which can engulf particulate matter, transforms into reproductive structures if under favorable conditions. The spores germinate producing mixamebe or flagellated cells , and when two of them merge, a new plasmodium is formed that becomes plurinucleate. In Plasmodiophoromycota , the plasmodial thallus within the host differentiates into zoosporeswith two smooth flagella, which at the end of the cycle become resistance spores (resting spores). The most important members of the group are Plasmodiophora brassicae , Spongospora subterranea and species of Polymyxa , parasites in the roots of many plants and important plant virus vectors.

Mushrooms similar to heteroconte algae

They are called pseudofungi because they differ in many characters from the true mushrooms and are classified in the Stramenopili taxon , together with heteroconte algae. These characters are:

Structural, presence of feathery flagella and dittiosomes biochemicals,

Presence of cellulose, sterols similar to phytosterols and particular tubulins

Genetic, have diploid nuclei and sequences similar to those of yellow and brown algae

On the other hand, they have lost the synthesis capacity of chlorophyll and have asexual spores with feathery flagella (zoospores).

The Oomycota phylum is the most numerous and includes aquatic and terrestrial species, often saprotrophic or parasitic or neurotrophic. Asexual reproduction generally involves the production of zoospores by cleavage in the sporangia and the swelling of the hypha with the formation of an oogony with one or more diploid nuclei that undergo meiosis and that are surrounded by distinct protoplasmic regions (oospheres). L ' anteridio , the male sexual organ, makes contact with the oogonio and merges with it. The male nuclei of antheridium migrate to the oogonium and fertilize the oospheres. A diploid zygote will form which will mature into resistant wall oospora.

The fungi feed for absorption, emit enzymes that break down the organic substance that can be absorbed. They also affect the bioavailability of metals since, by emitting enzymes, they also release H + (for reasons of charge compensation), thus lowering the ph.

As mycorrhizae they increase the absorption in plants of P, N (as well as H2O) as they manage to mobilize reserves containing these elements that the plant alone would not be able to exploit. Many plants thanks to mycorrhizae manage to take root even on unsuitable soils.

The fungi reproduce with enough fast times following sexual and asexual cycles, but also para-sexual phenomena, that there may be genetic variation in mitotic events. The biodiversity of fungi is almost as high as that of insects: the known insect species are around 2 and a half million and it is assumed that they are around 6 million; the mushroomson the other hand, around 80,000 of about 1.5 million are known to be discovered. The explanation of this high biodiversity can be given by the reproductive capacity and speed and by the adaptability. As far as the pH is concerned, we find species both at pH = 10 and at pH = 2;

For the temperature, there are thermophilic species whose spores resist up to 62 ° C and psychrophilic species whose spores resist up to -18 ° C; in water, in symbiosis with algae (lichens), in situations of lack of water they can also use poly-oils for the processes in which water is involved. Some species orient the structure that spreads the spores in relation to the light rays.

Oxygen: there are optional anaerobes and also obligatory (such as those that live in the rumen of herbivores.

The fungi are heterotrophic organisms; they feed by absorption through extracellular enzymatic production; they are eukaryotic organisms, not mobile; they can be single-celled (yeasts) and mycelia; in the mycelium there are transverse walls called septa which divide into secondary and primary.

Secondary septa are formed to block the escape of cytoplasm in case of mechanical damage, but also in response to the formation of reproductive structures, both in sexual and asexual reproduction. The material that is synthesized for reproductive structures does not go back to the ifa.

The fungimost primitive (Zygomycetes) do not have primary (or constitutive) septa. In general, the Zygomycetes have a greater ifa thickness (100-150 µm) than Ascom and Basidiomycetes (2-5 m). The larger diameter is connected to the fact that they do not have primary septa and that they have a smaller expansion of the mycelium. They are quite scarce as an enzyme kit and are therefore more likely to attack materials where simple sugars are present, in fact they are the first to attack the materials and reproduce very quickly; then the Basidiomycetes and Ascomycetes arrive which dismantle the more complex molecules, thus allowing the return of the Zygomycetes (which were in the dormant spore phase) which again have simple sugars available. Zygomycetes have poor competitive capacity, but have good growth and reproduction speeds. The Zygomycetes detach themselves from the water during the reproduction period (they do not have flagellated spores like Chitridiomycetes), but they also need a good percentage of humidity. They are typical of the woods and have no primary partitions.

Cellular septa - Ascomycetes and Basidiomycetes

In Ascomycetes and Basidiomycetes the primary septa have different morphology.

In Ascomycetes

They always present a central perforation that guarantees communication throughout the mycelium by passing cytoplasm, but also nuclei and cellular compartments and vesicles with enzymes: for this reason it is more correct to speak of mycelium and not of multicellularity, there is continuity inside of the whole mycelium. Poor substrates facilitate growth in length and when the mushroom finds nutrients, the branching increases a lot.

Hyphal anastomosis: hyphae of the same mycelium or hyphae of two different organisms of the same species can melt the walls and share cytoplasm and cellular compartments to promote growth, so as to allow a greater chance of survival for the new organism. The fungi are haploid organisms and thanks to the possibility of the passage of the nuclei and the possibility of fusion in some way compensates for the haploid.

To the right and left of the pore there are membranes, called Woronin bodies, which have the function, in case of damage, of closing the pore even before the secondary septum is formed.

In the single-celled forms of Ascomycetes(yeasts) the septum does not have this configuration. During reproduction, a gem is produced which grows to become the size of the mother cell, in the meantime mitosis occurs and the nucleus passes into the new cell. The primary septum does not have a single hole, but many small holes: the migration of the newly formed nucleus towards the daughter cell is faster than the synthesis of the septum, therefore when the nucleus migrates and the cell continues to grow the septum is not yet fully formed ; once formed, the septum allows the nucleus to remain in the new cell while allowing the exchange of cytoplasm. The new cell can remain attached to the mother cell going to form the pseudomycelium (yeasts).

In the Basidiomycetesthe

Septum has a central hole and has barrel-shaped enlargements on the sides (dolipore septum). Above and below the barrel there are membranes with numerous perforations (parentosomes) that cover the entrance of the septum and in case of mechanical damage they can prevent the escape of cytoplasm. The hole is smaller than the Ascomycetes , in fact the nuclei do not pass, but only cytoplasm and smaller organelles.

Depending on the environmental parameters, some species can pass from the unicellular to the mycelial form (dimorphism). All symbiotic mushrooms have this characteristic.

In human pathogens the passage from one form to another is dictated by the temperature and characteristics of the substances present in the substrate: they are in the yeast form inside the human organism as they can spread better, while outside they are in mycelial form. In the yeast form, the surface compared to the volume is reduced and this is an advantage against the attack of the antibodies (they are also less recognizable), for the pH and for osmotic issues.

When the fungus leaves the organism it passes to the mycelial form to feed better. Capsulated histoplasma at 37 ° C is in the yeast form and below 25 ° C is in the mycelial form. Plant pathogens are dimorphic in response to concentrations of sugar: at high concentrations of sugar (in the lymphatic vessels) they pass into the yeast form, then return to the mycelial form to colonize the organism and increase the contact surface. In some mushroomsthe transition from one form to another is dictated by oxygen. Many hyphae can modify their structure in order to increase the contact surface with the parasitic organism (appressorio). During contact, the appressory produces enzymes that degrade the host wall and then the infection stylet is introduced and the mycelium develops in the tissue, but it can also enter the cells. In mutualistic symbioses (arbuscular mycorrhizae) the penetration of hyphae into the cells occurs without damaging the cell membrane of the plant. Pathogens, on the other hand, during penetration cause significant damage to the cell.

In addition to the accessory, there may be lace-shaped hyphal modifications made up of three or four cells that have the characteristic of being able to react to tactical and chemical stimuli: in case of stimulus they swell trapping the prey (nematodes, rotifers, annelids and small insects) and emitting enzymes that digest it. This characteristic is present in the Zygomycetes which can also be used for biological control. In other cases the hyphae have structures that possess mucilaginous substances on the surface, the attracted prey remains stuck to the fungus and is digested. These are hyphal modifications that allow the use of substances that would otherwise not be accessible.

The wall in the apical area is thinner, as it is in this area that the hyphal elongation and the absorption of nutrients takes place. As we move away from the apical area there may be vacuoles which in the older areas can occupy a large part of the cell volume . Mitochondria are more numerous in the apical area because it is the most active area and therefore breathes more.

In the most distant areas the septum can be "closed", autolysis takes place, the wall thickens and becomes a resistance structure called chlamydospore which in suitable environmental conditions can give rise to another organism. From the apical area to those behind, therefore, the mitochondria, the nuclei decrease and the size of the vacuoles and the reserve substances (glycogen) that are scattered in the cytoplasm increase. In the apical area we have a series of vesicles (formed by the endoplasmic reticulum) which contain some lytic enzymes, other enzymes capable of synthesizing chitin: the vesicles once in contact with the wall merge, the lytic enzymes break the bonds of the wall, while other vesicles pour the precursors of chitin, in this way the wall lengthens and grows.

Cell wall

The mushroom wall consists mainly of chitin and other substances such as sugars which can be fibrillar in nature (chitin, and cellulose in the Oomycetes) and amorphous such as clifosans and R-glucans and S-glucans; the percentage of the latter two affects the rigidity of the wall . Melanin may also be present, which has the function of protecting it both from drying out and from UV rays.

The wall is made up of 4 layers (about 125 nm) with an alternation of polysaccharides and proteins, with the presence of lipids and, in the spore wall, also of sporopollenin which confers high resistance to acids. In some walls there are also alkanes that protect against dehydration.

Thicker walls are found in the sexual spores. The evolution has allowed that germination occurs only in response to certain factors: the spores of the truffles (reproductive structures of Ascomycetes), for example, do not germinate if they have not been eaten by animals.

The wall of the apical and sub-apical area is formed only by the two innermost layers.

First layer: chitin + protein matrix.

True mushrooms have ergosterol in the plasma membrane which is a typical animal sterol.

With the exception of Basidiomycetesmitosis is of the "closed" type, i.e. the nuclear membrane does not break.

In addition to lipids and glycogen (present in the form of rosettes in the cytoplasm) there may also be other reserve substances such as trehalose and poly-oils which have the function of helping resistance against dehydration: the OH groups of these poly-oils can be used for vital functions instead of the water molecule (such as lichens where the fungus receives the products of photosynthesis from the alga and in return offers protection and hydration); these substances are also involved in the maintenance of vital processes at low temperatures by lowering the freezing point.

In some yeasts chitin is present only in septa. The elongation in the yeast form is isodiametric (equal in all directions)

MUSHROOM KINGDOM
Phylum Chytridiomycota

It is the only phylum with flagellated elements, it includes just over 900 species. The members of this phylum have a single-cell thallus or consisting of a dichotomously branched cenocytic mycelium. In both cases rhizoids are present to anchor themselves to the substrate and to absorb nutrients. Some species have life cycles with haploid-diploid alternation. Sexual reproduction involves the fusion of male and female mobile gametes, although in the most advanced species the only mobile is male. Members of the Neocallimastigales order are the only known obligate anaerobic mushrooms. Some of these mushrooms are freshwater saprotrophs or in moist soils, while others are parasites of algae, nematodes or other fungi. Some are serious plant pathogens (Synchytrium endobioticum , agent of the black mange of the potato).

Phylum Glomeromycota

Contains about 150 species, they are among the most abundant in soils. They establish mutual symbioses with the roots of 80% of terrestrial plants. Their origin seems to be very ancient (about 460 million years ago) and it would have been the symbiosis with these mushrooms to allow the colonization of the emerged lands by the plants.

Phylum Zygomycota

They are the typical terrestrial mushrooms, generally saprotrophic on plant and animal residues of the soil. They act as pioneers since they are followed by species capable of exploiting the most complex undegraded substances. Some have powerful enzyme abilities and can break down chitin. The asexual spores are non-flagellated sporangiospores, while the sexual ones (zygospores) derive from the plasmogamy of specialized hyphae. Instead, sexual reproduction occurs when two individuals of different mating-types come into contact. If this happens, each mycelium produces short multi-nucleated branches (zygophores) which grow towards the branches of the other individual until they come into contact and dissolve the hyphal walls at the meeting point, followed by plasmogamy and pairing and subsequent fusion of the nuclei. This forms a zygosporangium , with diploid nuclei, which evolves into a zygospora , provided with a thick wall. Upon the arrival of favorable conditions, the zygospora germinates and undergoes meiosis, producing a sporangium capable of releasing new haploid spores.

Phylum Ascomycota

It is the largest group of fungi with 33,000 species united by the presence of axes, by sacciform structures that are home to cariogam and meiosis and by the formation of meiospores (ascospores). These mushrooms are, for the large number, very varied and able to live in very different environments. The thallus is generally mycelial septate, but it can also be yeast. The sporophores in this group are called ascomi and have a fertile region containing the aschi, and a sterile region. The germination of haploid spores originates septate mycelia which allow the passage of haploid nuclei. Asexually, this mycelium can produce conidia, while plasmogamy occurs between two differentiated organs of two compatible individuals: antheridium (male) and ascogonium (female).

The cariogam does not follow immediately, therefore dicarions are formed that will populate the hyphae originating from the ascogonium (ascogenic hyphae). These hyphae will fold into a crozier . At this point, the dicarions will divide by meiosis and the compartment where this occurs will turn into an ax. The four nuclei that came into being still divide by mitosis and separate from each other with the formation of walls, becoming ascospores . There may be cases in which there are more mitosis, therefore the number of ascospores can be much higher than 8.

Ascospores can finally be freed in different ways: actively expelled through a pore at the apex of the abs or even with the simple education of the wall itself.

Ascomi can be of different morphologies:

The cleistothecium is globular and without openings, and the axes inside are arranged in a disordered way;

The peritecium is piriform with an apical opening, the axes are arranged in a hymenium;

The apothecium is open, often cupped, with hymenium in the concave part, although some may be slightly different.

In yeast ascomycetes, asexual reproduction occurs by budding or fission.

Phylum Basidiomycota

It includes about 30,000 species united by the presence of the basidium, a structure similar both morphologically and functionally to the absus. The thallus is mycelial with a dolipore septum, but yeast-like basidiomycetes also exist. A basidiospore produces a haploid septate mycelium, called primary mycelium . Plasmogamy between two normal hyphae (somatogamia) occurs when two compatible primary mycelia meet . The compatible nuclei appear and the mycelium becomes secondary , this phase (called dicaryotic) can persist for years, with the presence of a dicarion in each hyphal compartment. At this point, regions of the secondary mycelium differentiate into sporophores of varying morphology.

The fertile region of the basidioma is organized in tubular or lamellar areas, covered with hymenium bearing the basidia. After cariogam and meiosis, the four haploid nuclei are brought to the outside of the basidium by offshoots. In this group there are many mushrooms of food interest (like some members of the Boletus genus , which also include porcini mushrooms, Boletus edulis , while others of the same genus are poisonous, Boletus satanas).

Anamorphic mushrooms (Deuteromycetes)

They reproduce mostly with the production of conidia, therefore asexually. They are similar to ascomycetes and basidiomycetes and in fact, in culture, they can produce sexual structures typical of these two groups. A connection is therefore created between the anamorphus (conidic phase) and the teleomorphic (sexual phase), which means that the deuteromycete represents the asexual phase of an ascomycete or basidiomycete.

The set of the two phases is called holomorphic and the mushroom is ordered based on the teleomorph. In addition to sexual structures, the ultrastructure of septa and DNA sequences lead to relationships with ascomycetes. Anamorphic mushrooms are among the most common: Penicillium , Trichoderma , Fusarium , Cladosporium , Alternaria , Aureobasidium are common soil saprotrophic mushrooms, while others are entomopathogenic (Beuveria , Metarhizium) or mycoparasites (Trichoderma).

The only organisms with cellulose cell walls are the Oomycetes which are not considered true fungi , but are inserted (considering the classification with 8 Realms) in the Chromists (or Stramenophyla), where some groups of brown algae are also inserted. The mushroom wall is responsible for recognition by other organisms.

Deuteromycetes refer to species of Ascomycetes and Basidiomycetes of which a sexual reproduction cycle is not recognized. Some of them are plant and animal parasites and cause diseases in higher animals.

Other mushrooms have been added to the Protist kingdom .

Molecular studies have shown that mushrooms are closer to the animal kingdom than to the vegetable one; they have glycogen as a reserve substance as in animals (in plants it is starch)

TYPES OF MUSHROOM AND THEIR CHARACTERISTICS

Edible mushrooms

Many of the more than 2,500 domestic mushroom species are edible - but far from all. The following overview presents the 30 most popular mushrooms in alphabetical order with their characteristic properties, classic locations and harvest time. If available, the list draws attention to the distinguishing features of toxic counterparts.

Birch mushroom (Leccinum scabrum)

Habit: 5 to 15 cm high, hemispherical, later flattened hat, 5 to 15 cm tall

Hat color: light gray-brown with reddish shimmer or yellowish nuances

Handle color: white, often with black net or black scales, similar to a birch trunk

Tubes (sponge): white on young mushrooms, later ocher yellow

Occurrence: in symbiosis under birch trees

Collection time: June to November

Flavor: mild to sour

Risk of confusion: unmistakable thanks to the characteristic handle

Birch mushroom (Leccinum scabrum)

Brätling (Lactarius volemus)

Growth habit: 10 to 12 cm high, 5 to 15 cm wide, flat hat with dent in the middle

Hat color: orange-red-brown to cinnamon, rarely bread yellow

Handle color: orange-brown

Slats: yellow with red discoloration at pressure points

Occurrence: on the edges of deciduous and coniferous forests

Collection time: July to October

Flavor: mild

Risk of confusion: unmistakable thanks to the pungent fish smell of herring

Special feature: strong milk flow occurs even with small injuries and turns the fingers brown

Brätling (Lactarius volemus)

Bronze boletus, black-capped boletus (Boletus aereus)

Growth habit: 15 cm high, 3 to 5 cm thick stem, hemispherical, up to 25 cm tall hat, velvety skin

Hat color: black to ocher-brown

Handle color: creamy yellow to creamy white

Tubes (sponge): white

Occurrence: under chestnuts and oaks

Collection time: July to October

Flavor: nutty and mild

Risk of confusion: none, resembles other boletus mushrooms, all of which are edible

Special feature: mushroom of the year 2008, one of the most impressive, edible mushrooms on home soil

Bronze boletus, black-capped boletus (Boletus aereus)

Thin meat anise mushroom (Agaricus silvicola)

Form of growth: 8 to 10 cm high, thin hat with 5 to 10 cm diameter, first bell-shaped, later flat screened

Hat color: sulfur yellow to cream with yellow spots

Stem: white, hollow inside, slim, 1 to 1.5 cm thick, bulbous foot

Slats: pink, brown with age

Occurrence: deciduous and coniferous forests

Collection time: June to October

Risk of confusion: poisonous tuberous agaric with white lamellae

Special feature: intense scent of anise or almonds

Thin meat anise mushroom (Agaricus silvicola)

Noble stimulus, blood stimulus (Lactarius deliciosus)

Growth habit: 3 to 7 cm high, 5 to 10 cm wide, with a hollow foot when old

Hat: first flat, later funnel-shaped to rolled up, brick-orange with shades, silvery stripes

Handle color: orange

Slats: ocher yellow to light orange

Occurrence: pine forests

Collection time: August to October

Flavor: mild, fruity

Risk of confusion: irritant salmon, which is not poisonous but has an extremely bitter taste

Special feature: orange milk turns the urine red after consumption, which is harmless

Noble stimulus, blood stimulus (Lactarius deliciosus)

Noble stimulus, blood stimulus (Lactarius deliciosus)

Bottle Dustling (Lycoperdon perlatum)

Growth habit: prickly, white shape, 3 to 10 cm tall, no pronounced hat

Handle color: white for young mushrooms, brown for older specimens

Occurrence: coniferous forests, rarely in deciduous forests

Collection time: June to October

Risk of confusion: poisonous potato biscuit with yellow-brown warts and black meat inside

Special feature: brown meat from older mushrooms is inedible

Bottle Dustling (Lycoperdon perlatum)

Bottle Dustling (Lycoperdon perlatum)

Female pigeon (Russula cyanoxantha)

Habit: 5 to 15 cm wide, flat hat

Hat color: violet-green, more rarely ocher-yellow

Stem: 4 to 10 cm high, 1 to 4 cm thick, whitish with a purple hue

Slats: white

Occurrence: beech forests and in mixed forests

Collection time: July to October

Flavor: pleasantly mild

Risk of confusion: with other pigeons whose bitter taste can numb the tongue

Special feature: mushroom of the year 1997

Female pigeon (Russula cyanoxantha)

Female pigeon (Russula cyanoxantha)

European Boletus (Suillus grevillei)

Growth habit: first hemispherical, later flat hat, 5 to 15 cm wide

Hat color: gold-yellow to orange-brown

Stem: yellowish, 4 to 10 cm high and 0.5 to 2 cm wide, white skin ring

Tubes (sponge): yellowish, brown when old

Occurrence: among larches in symbiotic community

Collection time: July to October

Flavor: spicy-mild, with tender meat

Risk of confusion: none

European Boletus (Suillus grevillei)

European Boletus (Suillus grevillei)

Hawk mushroom, venison mushroom (Sarcodon imbricatus)

Habit: 5 to 8 cm high with a hat up to 15 cm wide, spiky underneath

Hat: scaly and light brown like the plumage of hawk and sparrow hawk, with a sunken middle

Stem: 2 to 5 cm wide, initially white, later turns brown from the base

Slats: light gray, spiky

Occurrence: in beech and spruce forests, often in closed witch rings

Collection time: June to November

Flavor: mild to nutty-spicy

Danger of confusion: due to the spiky underside of the hat, easy to identify even by laypersons

Special feature: mushroom of the year 1996

Hawk mushroom, venison mushroom (Sarcodon imbricatus)

Hawk mushroom, venison mushroom (Sarcodon imbricatus)

Hornbeam bolete (Leccinum carpini)

Growth habit: wrinkled hat with a diameter of up to 12 cm

Hat color: light brown to dark brown

Handle color: greyish with dark scales

Tubes: bulge-like bulge around the stem gray-white, later gray-green

Occurrence: among hornbeams and hornbeam hedges

Collection time: July to October

Flavor: pleasantly mild

Risk of confusion: none, resembles the edible birch mushroom

Special feature: after being cut, the meat turns purple to black, which does not affect the taste

Hornbeam bolete (Leccinum carpini)

Hornbeam bolete (Leccinum carpini)

Fall Trumpet (Craterellus cornucopioides)

Growth habit: funnel-shaped, hollow fruiting body, 2 to 6 cm wide, 3 to 12 cm tall

Hat color: black inside, brown-black outside with folded edge

Handle color: gray

Occurrence: mainly under beech trees, otherwise in beech and fir forests

Collection time: August to November

Flavor: mild

Danger of confusion: Gray performance with 1-6 cm wide, gray-brown funnel hat

Fall Trumpet (Craterellus cornucopioides)

Fall Trumpet (Craterellus cornucopioides)

Judas ear (Auricularia)

Form of growth: 4 to 10 cm wide, elastic fruiting body reminds of an auricle

Fruit body color: reddish brown, olive brown to greyish violet

Flesh color: reddish brown

Occurrence: mainly on the wood of the black elder and on other deciduous trees

Collection time: all year round

Flavor: very mild

Risk of confusion: unmistakable due to the bizarre shape

Special feature: Mushroom of the year 2017, contracts when it is dry and swells when it is wet

Judas ear (Auricularia)

Judas ear (Auricularia)

Curled hen, curled hen (Sparassis crispa)

Growth habit: frizzy fruiting bodies, 10 to 40 cm in diameter, resembles a bath sponge

Color fruiting body: yellowish to light brown

Flesh color: white-yellow

Stalk: thick fleshy base, reminiscent of a cauliflower drink

Occurrence: pines

Collection time: July to November

Taste: very tasty, mild nutty, the lighter the flesh, the more delicate

Risk of confusion: none, the similar looking oak hen is also edible

Curled hen, curled hen (Sparassis crispa)

Curled hen, curled hen (Sparassis crispa)

May mushroom (Calocybe gambosa)

Habit: 5 to 8 cm high, 3 to 10 cm wide, initially closed, later spread hat, rolled up edge

Hat color: cream-white

Handle color: cream-white

Slats: cream-white

Occurrence: on forest edges

Collection time: May and June

Flavor: mild

Risk of confusion: highly toxic brick-red crack fungus, optical double, whose fruit-like scent serves as the most important distinguishing feature

Special feature: more intense, than obtrusive, mealy smell that disappears when cooking

May mushroom (Calocybe gambosa)

May mushroom (Calocybe gambosa)

Chestnut Boletus (Xerocomus badius)

Growth habit: 12 cm high, 3 cm wide, semicircular

Hat color: brownish, matt, greasy when wet

Handle color: yellow-brown

Tubes (sponge): cream-yellow, discolored bluish with slight pressure

Occurrence: Spruce and pine forests

Collection time: June to November

Flavor: nutty, mildly aromatic

Risk of confusion with toadstools: none

Chestnut Boletus (Xerocomus badius)

Chestnut Boletus (Xerocomus badius)

Black-headed milkling (Lactarius lignyotus)

Growth habit: hat with rolled edge, 2 to 6 cm wide, grooved stem

Hat color: black-brown to soot-colored

Handle color: black-brown

Flesh color: white

Slats: white

Occurrence: spruce forests

Collection time: August to October

Taste: mild, but the milk can taste bitter

Risk of confusion: pitch-black milkling, unground, darker stem, hunched over, causes stomach upset

Black-headed milkling (Lactarius lignyotus)

Black-headed milkling (Lactarius lignyotus)

Monk's head (Clitocybe geotropa)

Form of growth: 8 to 15 cm tall, with up to 25 cm wide, spreading hat and wart-like hump

Hat color: cream to beige, brownish in age

Handle color: cream to beige

Slats: white to off-white, down the stem

Occurrence: deciduous and coniferous forests

Collection time: September to November

Risk of confusion: highly poisonous white varnished funnel, only half the size and with no hump on the hat

Special feature: gathers up to 800 m tall witch rings

Monk's head (Clitocybe geotropa)

Monk's head (Clitocybe geotropa)

Pearl mushroom (Amanita rubescens)

Growth habit: spherical, later screened hat, up to 15 cm tall

Hat color: light brown, set with pink-gray pearl flakes

Handle color: white with a reddish tinge, fluted ring

Slats: white, spotted pink when aged

Occurrence: deciduous and coniferous forests

Collection time: June to October

Taste: mild, sweet, raw poisonous

Danger of confusion: panther mushroom, very poisonous, white stem without red shimmer, hat with smaller pearl flakes

Special feature: the pearl-like flakes disappear from the hat when it rains

Pearl mushroom (Amanita rubescens)

Pearl mushroom (Amanita rubescens)

Chanterelle (Cantharellus cibarius)

Growth habit: hemispherical, short-stemmed, 3 to 8 cm tall

Hat: bright yellow, 3 to 12 cm in size, changing shapes from deep recessed to arched to funnel-shaped

Stem: thin, grooved and yellow

Slats: yellowish and forked

Occurrence: coniferous forests, mainly in nests under pines

Collection time: June to October

Flavor: spicy to slightly peppery

Risk of confusion: none, resembles the false chanterelle, which tastes bland

Special feature: is under nature protection and may only be collected for personal use

Chanterelle (Cantharellus cibarius)

Chanterelle (Cantharellus cibarius)

Giant umbrella (Macrolepiota procera)

Growth habit: 15 to 40 cm high, young hat round, screened up to 40 cm tall, with a slight hump in the middle

Hat color: cream-colored with dark scales

Handle: slim, hollow, with a sliding ring, gray-brown

Slats: creamy white, later brownish

Occurrence: in clearings and on the edges of deciduous forests, in parks and in cemeteries

Collection time: June to November

Taste: nutty to mild, especially the ring

Risk of confusion: poison saffron umbrella, 5 to 30 cm tall, light-colored hat with dark spots, causes stomach upset

Special feature: edible mushroom of the year in 2017, still raw poisonous

Giant umbrella (Macrolepiota procera)

Giant umbrella (Macrolepiota procera)

Sheep champignon (Agaricus arvensis)

Habit: 5 to 15 cm high, hemispherical to slightly arched, very meaty hat, up to 15 cm in diameter

Hat color: cream-colored, smooth surface

Handle color: cream to dirty yellow

Slats: gray-white to pink and brownish

Occurrence: meadows, parks, pastures

Collection time: May to October

Flavor: tasty, mildly pleasant

Danger of confusion: highly toxic carbol mushroom, very similar in appearance, carbol-like, unpleasant smell is the most important distinguishing feature; deadly poisonous spring tuber agaric, recognizable by pure white lamellae, therefore please exercise extreme caution with young mushrooms that still thrive with white lamellae

Sheep champignon (Agaricus arvensis)

Sheep champignon (Agaricus arvensis)

Crested Inkling (Coprinus comatus)

Growth habit: 10 to 25 cm high with a cylindrical hat

Hat: 5 to 10 cm high, 3 to 6 cm wide, white with brown scales

Handle color: white

Slats: white

Occurrence: rubble sites, fields and roadsides, meadows, rarely in forests

Collection time: May to November

Taste: mild, delicate, but only with young mushrooms

Risk of confusion: unmistakable

Special feature: old specimens melt away, like dark ink

Crested Inkling (Coprinus comatus)

Crested Inkling (Coprinus comatus)

Pig's ear, purple cuckoo (Gomphus clavatus)

Growth habit: 2 to 8 cm high, 4 to 8 cm wide, often intertwined, funnel-shaped fruiting bodies, wavy edge

Fruit color: flesh-colored to purple

Last on the outside: flesh-colored, longitudinally veined, forked

Occurrence: beech and spruce forests

Collection time: August to October

Danger of confusion: due to the unusual shape, unmistakable even for laypeople

Special feature: mushroom of the year 1998, is under nature protection, collecting for own use is allowed

Pig's ear, purple cuckoo (Gomphus clavatus)

Pig's ear, purple cuckoo (Gomphus clavatus)

Food morel (Morchella esculenta)

Growth habit: 10 to 30 cm in size, oval to egg-shaped with round, honeycomb-shaped ribs

Hat color: light beige to brownish

Handle color: white, lighter than the hat

Occurrence: deciduous forests, parks

Collection time: April to May

Danger of confusion: none, resembles the less tasty pointed morels with a pointed, net-like hat

Special feature: is under nature protection, collecting for personal use is allowed

Food morel (Morchella esculenta)

Food morel (Morchella esculenta)

Boletus, spruce boletus, male mushroom (Boletus edulis)

Growth habit: thick tube, rounded, domed hat

Hat color: brown

Stem: pale brown, thick-bellied, barrel to roller-shaped, with a light mesh pattern

Tubes (sponge): with increasing age from white to yellowish to green

Occurrence: coniferous and deciduous forests

Collection time: June to October

Flavor: nutty

Danger of confusion: none, resembles non-toxic, bitter, light brown gall blanks with a pink underside

Special feature: is under nature protection, picking for own use is allowed

Boletus, spruce boletus, male mushroom (Boletus edulis)

Boletus, spruce boletus, male mushroom (Boletus edulis)

Stick sponges (Kuehneromyces mutabilis)

Growth habit: 5 cm high, 4 cm wide,

Hat color: brown, damp shiny

Stem: a ring separates the smooth, upper, cream-colored half from the scaly, dark, lower half of the stalk

Slats: cream white

Occurrence: deciduous forests

Collection time: April to November

Flavor: spicy

Danger of confusion: poison-dove with a smooth, cream-colored to brown stem as the only distinguishing feature

Stick sponges, Kuehneromyces mutabilis

Violet lacquer funnel (Laccaria amethystina)

Habit: 4 to 10 cm high, 2 to 6 cm tall, flat to domed hat

Hat color: when wet: intense violet, the drier the paler

Handle color: violet, fibrous

Slats: purple, strikingly far apart

Occurrence: deciduous and fir forests and in parks

Collection time: June to November

Taste: rather inconspicuous, but gives mushroom dishes a colorful peppery taste

Risk of confusion: none, resembles the following edible purple ruby knight

Violet lacquer funnel (Laccaria amethystina)

Violet lacquer funnel (Laccaria amethystina)

Purple Ruby Knight (Lepista nuda)

Habit: 5 to 15 cm high, smooth, shiny, domed to funnel-shaped hat, with a flat, rolled-up edge when old

Hat color: bright violet, brownish with age

Stem: purple and bulbous with silvery-white longitudinal fibers

Slats: purple

Occurrence: beech and spruce forests, sometimes on meadows or on compost heaps

Collection time: September to November

Risk of confusion: none

Purple Ruby Knight (Lepista nuda)

Purple Ruby Knight (Lepista nuda)

Meadow mushroom (Agaricus campestris)

Growth habit: 10 cm tall, initially spherical, later flat hat, similar to the mushrooms in the shop

Hat color: gray-brown to white

Handle color: white

Slats: gray-pink, later dark gray to black

Occurrence: meadows, green areas, paddocks, grasslands

Collection time: May to November

Risk of confusion: poisonous carbol mushroom, unpleasant smell, chrome-yellow flesh in the base of the stem; poisonous white tuber agaric with white lamella

Meadow mushroom (Agaricus campestris)

Meadow mushroom (Agaricus campestris)

Winter turnip, velvet foot turnip (Flammulina velutipes)

Growth habit: 2 to 10 cm wide hat, thin-fleshed, with a greasy surface

Hat color: honey-yellow to reddish-brown with a dark center

Stem: ringless, velvety to felty, 3 to 8 cm long, yellow-brown above, dark brown below

Slats: white to pale yellow

Occurrence: on trunks and stumps of living or dead trees

Collection time: September to April

Flavor: tasty edible mushroom

Danger of confusion: poisonous pet with a smooth handle as the most important distinguishing feature

Special feature: even grows through a blanket of snow

MUSHROOM LIFE CYCLE

When I started the mushroom journey (and obsession), understanding the basic life cycle helped me in many ways. First, I had to learn the terminology I usually ignored, which made me a better identifier and mushroom picker. In the end, it made me appreciate nature planning - not just the mushroom, but the mushroom's connection to everything that surrounds it.

Let's take a (very) basic look at the mushroom life cycle and start with the most recognizable part of the mushroom - the fruit body - with the button.

Kind of mushroom

The mushroom: the fruit body of the mushroom. Just as a tree yields fruits that contain seeds for growth, the "fruit" of the fungus is the fungus, and the fungus produces spores (such as seeds). (To learn more about the difference between a mushroom and a mushroom, click here.)

Spore: fungal growth units, microscopic and usually unicellular, but not always, similar to flower seed. Fungi can release their spores in various ways and ensure reproduction. This is one of the most interesting aspects of the mushroom. And as diverse as fungi, just as spores are in shape and size.

You may have heard the terms Basidiomycetes and Ascomycetes. These are two dominant subgroups of fungi that differ from the type of spore-producing and releasing structure.

Basidiomycetes, think of the typical gill mushroom, stem and cap. Their basidiomycetes spores produce their spores on nodular structures called "basidium". The spores are formed on the outside of the small mountains at the end of the base. These spores usually produce four at the end of each basidium and are called basidiospores.

Ascomycetes are fungi of ca. 75%. They consist of things like shellfish, truffles and cookies Their Ascomycetes spores are produced inside the long, sack-like structures called "Asci" (plural "Ascus"). The ascus typically produces eight spores in each bag-like structure, and the spores are called ascospores. Although the two spore-retaining structures are different, the goal is the same: to produce and project spores for new fungi.

In most cases, thousands and thousands of spores are released from a single fungus. In fact, some species claim to have more than 30 billion wounds a day! Some spores may begin the reproductive process on the same day after liberation, while others must go through a certain process before they can germinate. There are many ways to reproduce the spore, so far all of the reproductive processes are not fully understood, and it would take too long for someone to understand and explain everything. For simplicity, we will talk about the process of sexual reproduction through which the fungus undergoes.

When a spore germinates, the spore forms a filamentous structure called a separator. It branches off as the dash grows and eventually attaches to a separate dash from a compatible spore. As far as the reproduction of sexual fungi is concerned, there are not necessarily "male" and "female" structures, but "mating types", sometimes called "positive" and "negative" or "plus" and "minus". When two compatible hyphae meet, genetic information is exchanged. With this exchange, the hypha has all the information they need to reproduce. The associated hyphae branch rapidly and this branched network is called mycelium.

Dash: Plural, hyphae. The filiform fungal cell, which together forms the mycelium.

Mycelium: The vegetative part of fungi, which is composed of a complex hyphae network and often resembles the "root" of a fungus in which nutrient and chemical exchanges take place.

The mycelium can grow far and wide, be connected over long distances and form an incredibly large network under our feet. You will hear the term "mycelium mat" referring to an area of the soil that is usually just below the surface and interwoven with the mycelium -

sometimes dense enough for dirt and duff to adhere to the strength of the mycelium.

When the ideal temperature, humidity and nutrient factors come together, the mycelium begins to develop fungi. The mycelium wraps up tightly, twists and condenses to form the so-called hypothalamus, which turns into an easily recognizable form of the fungus. This small mass of mycelium that looks like a baby mushroom is called a primordium.

Primordium: The earliest recognizable stage of the body in which it develops, in this case a fungus.

The Primordia continues to develop and the baby mushroom begins to grow. Some mushrooms are predefined, others undefined - what does that mean? A predetermined fungus develops with all necessary parts (e.g. stem, cap, gills) in the earliest stages of life, that is, preformed, and when it is harmful at a young age, these defects appear in the mature fungus. The shape of the undetermined mushroom at the time of ripening has not yet been determined. During ripening, this mushroom determines its final shape based on its environment. An undetermined mushroom can be inhibited if it matures like a branch and grows around it, even if it is caught. If they are injured, they will recover without harm.

When the mushroom matures, the spores are produced and released, and the cycle continues. For those who follow my Instagram page, posts are often shown about whether the fungus creatively protects its spores at the right time and in the right way. We've already talked about what happens to the spores, but what the fungus is after the spores are released (if a hungry hunter didn't collect them for the table).

Many insects, worms, and other criminals begin to eat the decaying fungus until it is in an almost unrecognizable state. The spores of the mushroom germinate and form a dense mycelium network over the dead mushroom that remains on the meat. Then the mycelium retreats into the dirt and duff where we think it is hidden, but life spreads out carefully.

In summary: the spores germinate, the hyphae combine and form a mycelium. The mycelium forms a primordium, which becomes spore-releasing fungi, which are decomposed by insects and mycelium consumption.

The terminology is:

Ascomycete: Mushrooms whose spores grow in long, sack-like structures, which are referred to as "Asci" (plural "Ascus"). The ascus typically produces eight spores in the structure of each sack, referred to as spores.

Basidiomycetes: Fungi that grow their spores on lump-like structures called "basidium" grow spores on the ends of small mountains from the outside. Typically four are produced at the end of each base, these spores are called basidiospores.

Hyphe: plural, hyphae. A fibrous fungal cell that together forms a mycelium called mycelium.

Mushroom: The fruity body of a mushroom. Just as a tree produces fruit that contains seeds for growth, mushrooms are the "fruit" of the mushroom, and mushrooms produce spores (like seeds). (To learn more about the difference between mushrooms and mushrooms, click here.)

Mycelium: The vegetative part of mushrooms, which consists of a complex hyphene network, often resembles the "root" of the mushroom, where nutrients and chemicals are exchanged.

Predefined mushroom: The mushroom with all the necessary components (stem, cap, gill) is formed at an early stage of life, i.e. it is "preformed" and develops into a mushroom when damaged at a young age.

Primordium: The earliest recognizable stage of the body in which it develops, in this case a fungus. Unidentified mushroom: A mushroom whose shape is not yet ripe. During ripening, this mushroom determines its final shape based on its environment. An undetermined mushroom can encounter obstacles because it matures and grows like a branch, even immersed in it. If they are injured, they will recover without harm.

Spore: Fungus growth units, microscopic and generally unicellular, but not always similar to a flowering plant as seeds.

MUSHROOM CULTIVATION: How to grow mushroom yourself

You can easily grow many types of mushrooms at home. In principle, own mushroom cultures on straw, wood or pre-made mushroom substrates are possible. But in the beginning there are the mushroom brood - mushroom spores or the living mushroom culture, which is on a carrier material. Mushroom broods come in various forms. In brood kernels, the mycelium, i.e. the mushroom network, has spun its threads around and into cereal or millet grains. The organic nutrients of the grains serve as a food basis for the mycelium. Grain brood can be mixed very well with substrate and simply packed in cans or bags in this form.

Fermented, mixed straw flour, chopped straw or even sawdust serve as the basis for the substrate brood. This brood is ideal for spitting straw bales or soaked straw pellets. The mass is simply broken into pieces the size of a nut. Chopsticks or dowel brood are traditional beech wood dowels from the hardware store, which are, however, completely permeated by the mushroom mycelium. The chopstick brood is ideal, for example, for spiking trunks or straw bales.

A mushroom brood can be stored at temperatures between two and twelve degrees Celsius for up to twelve months before it has to be processed. The lower the temperature, the longer the shelf life. Before contact with the mushroom brood, you should either wash your hands thoroughly or wear sterile disposable gloves so that no bacteria or mold spores stick to your hands. If one infects the brood with the adhering pests, the whole culture can die.

After successful vaccination of the carrier material, a white fluff is initially visible on the surface. This is the sign that the mycelium has already completely grown through the earth or the trunk. In the next stage, small white nodules appear, so-called primordia - mushrooms in the absolute mini format. But within a few days the primordia ripen into real mushrooms. This process is known as fructification (fruit formation): the visible mushrooms that can later be eaten are actually only the fruiting bodies of the mushroom network. They carry the spores over which the mushrooms sow themselves.

Grow mushrooms with coffee grounds

When growing mushrooms, a special substrate based on straw or bark mulch is usually used as the breeding groundor grain used. Herb mushrooms, lime mushrooms or pioppino can also be grown on coffee grounds that you have collected yourself. The mushroom brood is first crumbled into millimeter-sized pieces and mixed with dried coffee powder. Then you put everything in the growing pot, cover it and keep the mushroom substrate moist. After two to four weeks, when the white-gray mushroom threads (mycelium) have completely grown through the substrate, the cover is removed. The mushrooms appear in several batches. After about six harvest waves, the nutrients contained in the coffee grounds are used up.

Tip: As soon as the temperature rises above ten degrees Celsius, you can take the mushroom culture out of the pot and sink it into the ground in a shady place in the garden.

Mushrooms Pioppino, a noble mushroom with a velvety hood, which is particularly popular in Italy, can be grown on coffee grounds

Ready-made crops for growing mushrooms in the house

Oyster mushrooms should always be grown as ready cultures according to the enclosed instructions. Usually a fully grown substrate block is delivered. Without any intervention, the first harvest is often possible after just a few days. Reason: During transportation, the block was exposed to vibrations that stimulated fungal growth. Now it is time to store the substrate bale in a humid room or to bring the right humidity with a film. The block should always be kept moist. The excess water can be collected in a bowl. Don't forget the air holes, because they also promote growth. The optimal temperature is 18 to 25 degrees Celsius.

Oyster mushrooms

Professional mushroom growers also use substrates

If the mushroom culture feels good, the first fruiting bodies begin to form in the air holes. Depending on the type of mushroom, the bag is cut down to the substrate. As soon as the mushrooms have reached a size of eight to twelve centimeters, they can be carefully unscrewed or cut off with a knife. If possible, without leaving a stump, otherwise putrefactive bacteria can penetrate here. After the harvest there is a rest period of up to 20 days. After four to five harvest phases, the substrate is exhausted and can be added to organic waste or compost.

Mushroom growing in substrate

The white fruiting bodies of what is probably the most popular cultivated mushroom are lined up. The mushroom substrate has grown excellently in the growing tray.

Mushrooms are supplied as ready-grown cultures as a streaky substrate. An additional bag contains the covering soil. The substrate is spread out in a growing tray and covered with the supplied soil. The vessel is then covered with a transparent plastic hood. If you do not have a growing tray, you can also line a small wooden box or any other container with foil and put the substrate and the covering soil on it. Now it is important to keep everything moist. The mushroom culture requires temperatures between 12 and 20 degrees Celsius. The wooden boxes are best covered with a film first. As soon as the primordia show up, the cover has to be removed, because now the mushrooms need fresh air to thrive.

Grow mushrooms outdoors

There are different cultivation methods for growing mushrooms outdoors, all of which have their advantages and disadvantages. Which one is the right one mainly depends on the particular type of mushroom.

High-pressure pressed straw bales are particularly suitable for oyster mushrooms and brown caps. In April or May, the bales are completely soaked in clear rainwater for two days in a rain barrel or an old bathtub, after which you let them drain for a day. Then a stick or grain brood is brought out: Drill holes in the straw bale with a planting wood at a distance of 20 centimeters and insert the brood into it. After the straw bale has been completely peppered, it is covered with a film. It increases the humidity and offers the mycelium optimal growth conditions. After about five to six weeks between 20 and 25 degrees Celsius, the mycelium has completely penetrated the bale. Important for the coming weeks: The straw must always be kept moist, but must not be wet. With good care, a fine web will appear after three weeks. The first mushroom harvest takes place three weeks later. Depending on the weather, this method can be used to harvest up to six kilograms of mushrooms - always in cycles of three to four weeks. Then the straw is exhausted as a food source for the mycelium and migrates to the compost.

Instead of a straw bale, you can also use straw pellets. Oyster mushrooms, brown cap, herb mushrooms, lime mushrooms, pink mushrooms and crayfish are very suitable for this type of mushroom culture. Moisten the straw pellets in the bucket with water so that they can swell. Then you have to mix in a grain or substrate brood, fill the mass in plastic bags and bind. Insert a few air holes so that the culture can breathe. The whole thing is stored in a shady place at about 15 degrees Celsius. The white mycelium appears three weeks later and the plastic film can be removed. The culture needs a bright but not full sun place for further growth

Oyster mushrooms cultivated with straw

You can also cultivate oyster mushrooms and pink Seitling in pots filled with straw pellets, for example on the terraceor on the balcony. Moisten the straw pellets in a container with warm water and let them swell for three to four hours until the pellets fall apart. Then add water again until a porridge is formed. Spread a substrate brood evenly over the porridge and mix in. The mass is then poured into the pots, covered with foil and left to stand for about six to ten days so that the mixture can ferment (ferment). Through the holes in the flower pots, the water can easily drain into the coasters. Make sure that the latter is emptied regularly so that there is no waterlogging. After four to six weeks, the first of a total of three to four harvest cycles begins.

Grow mushrooms on logs

Mushroom cultivation on wood is particularly productive because with this cultivation method you can harvest mushrooms for up to seven years after vaccination. Oyster and shiitake mushrooms are ideal for this method. They extract the necessary nutrients from the rotting wood until the wood is completely drained. The wood should have been cut from four weeks to five months before vaccination. In contrast to the Shiitake, the oyster mushroom requires earth contact via the log, which is why the lower end of the trunk is buried about 20 centimeters deep in a shady place that is as free of snails as possible. Stick or grain brood is best suited for inoculation. Before this, the trunk or branch, which is about one meter long, must be well watered in a rain barrel or a tub.

Oyster mushrooms in the tree trunk

Tree trunks are a rich source for mushroom lovers

There are various methods for vaccinating logs. With the head vaccination, the stem is placed vertically and a disc (approximately five to ten millimeters thick) is cut off with the saw. The resulting cut surface is covered with grain brood and the previously separated slice is replaced and fixed in the center with a nail. Then the interface is sealed on the side with an adhesive strip. So that it cannot come loose even in damp conditions, it is also tied with push pins. To inoculate the cut, use the tip of the chain saw to cut several notches about five centimeters deep into the trunk at intervals of 15 to 20 centimeters. They are also filled with grain brood, pressed briefly and fixed with adhesive tape and thumbtacks as described above.

Borehole vaccination is also a very popular method. To do this, drill approximately three centimeters deep holes in wooden dowel thickness around the trunk at a distance of about 20 centimeters in a spiral arrangement. Then you push a stick brood into each borehole and seal the hole with tape or candle wax. Instead, you can simply wrap the entire trunk in plastic film. In a shady place at up to 25 degrees Celsius, mycelium permeates softwood such as poplar or birch after about six months. With hardwood, this takes about ten to twelve months. If black or brown mold appears, the vaccination has failed.

A particularly quick and easy method is the mushroom culture on plywood panels. Let two plywood sheets made of poplar wood soak in water and coat a plate with grain brood. Place the second plate on top and fix both at all four corners with wood screws. After a short storage in the foil bag, the plates are completely covered by the mycelium and there is nothing left to harvest soon.

Other cultivation methods and storage

If you have a small forest plot or a shady wooded area with humus-rich soil, you can also cultivate your mushrooms directly in the soil. This method is suitable for oyster mushrooms, brown cap, herb mushrooms, stick sponges and Tuscan mushrooms. As with the culture, let straw pellets swell in warm water. Then dig a 50 x 50 cm and 15 cm deep hole at the designated place in the garden and pour half of the swollen mass into it. Spread the brood (substrate, grain or stick brood) evenly over it and pour the other half of the straw pellet pulp over it. Press the mixture gently and cover it with a layer of earth two fingers wide.

Oyster mushrooms harvest

Oyster mushrooms in the forest garden just before harvest

Oyster, lime mushroom and shiitake are available as ready cultures for the field.

Growing a ready culture outdoors works similarly to indoors. The finished culture is usually delivered in a plastic bag. You simply have to place it in a shady place in the garden according to the instructions and always keep it moist. After two to four weeks, the first harvest wave follows, followed by a rest period of up to four weeks. A total of up to five batches are possible.

Store mushrooms correctly

Fresh mushrooms can easily be stored in the refrigerator for four to five days. If they have become a little dry during this time, no problem: sprayed with a little lukewarm water, they are quickly plump and fresh again. If there is excess mushroom, you have the option of freezing or drying mushrooms . For freezing, you can choose between the variant "in one piece" or "sliced".

Dried mushrooms

Fresh out of the oven: dried mushroom

For drying, place the mushrooms in slices on a baking sheet lined with baking paper. Preheat the oven to 50 ° C and let the mushrooms dry with the oven door open until any liquid has escaped. You can achieve the same result with a conventional drying machine. For subsequent use, the mushrooms are simply soaked in a little water until they have regained their typical consistency

Ready-made mushroom cultures for home and garden

One can obtain ready-made mushroom cultures for mushroom cultivation at home in shops or on the Internet . There are different types, from the pure culture in the Petri dish to the ready-to-spread or ready-to-put mushroom brood to the completely prepared ready culture in the practical growing box.

Pure mushroom mycelium

Petri bowl

You can grow your mushroom culture yourself.

Pure mushroom cultures are usually offered in petri dishes. It is the mushroom mycelium in the "basic stage" . Such pure cultures can also be produced by adding fungal spores of the desired type of fungus to a nutrient solution and growing under humid conditions. However, this process must be completely sterile, otherwise there is a high probability that mold spores will creep in and destroy the culture.

Pure cultures are only the preliminary stage to the "mushroom brood", which is ultimately used to "inoculate" the fungus into a substrate. If you put the culture in a sterile, moist organic substrate (e.g. a bottle with boiled cereal grains), the mycelium grows through the grains and after a while you can inoculate a prepared piece of wood, a straw bale or similar suitable carrier material with it. Because a lot of technical equipment, know-how and time is required for this variant, it is better to skip the first step and buy ready-to-eat mushroom brood.

Mushroom brood, mycelium patches and sticks

Mushroom brood, mycelium sticks or patches are ready-to-use mushroom cultures . If these mushroom cultures are placed in a suitable substrate and stored in a moist, warm condition, after a few weeks the mycelium penetrates the carrier material, from which the desired fruiting bodies finally sprout.

There are various options for the vaccination process: straw, logs or potting soil with a high organic content can be used . Other organic materials such as rabbit litter or bark mulch are also conceivable. The most important thing is that the substrate is kept permanently moist and is always stored between at least 5, better still 15 and 25 degrees. The ideal breeding temperatures may vary depending on the type of mushroom.

Sometimes the mushroom brood is already supplied with a suitable covering material, which the mushroom mycelium grows through during the breeding phase. This is standard for finished crops; they are even easier to use.

Ready crops

Finished mushroom cultures come in packages with a ready-to-use or already grown substrate. The stage of development of the mycelium determines how long it takes before the first mushrooms can be harvested after opening the culture. You simply follow the instructions and often have the first mushrooms to harvest after a short time. Usually the container is first opened and the culture is moistened and / or covered with substrate and then closed again to keep the moisture.

Depending on the culture, the whole thing remains permanently covered like a mini greenhouse or can be opened or cut after a certain ripening period. "Ripe" is the substrate bale when the white mushroom mycelium has completely penetrated it. Due to the influence of oxygen, mushrooms begin to sprout under the film or through the openings and can be harvested in several harvesting cycles over weeks. After three to five harvest phases, the entire package can finally be disposed of with household waste.

Why does the mushroom culture need to be covered?

Forest trees constantly release moisture into the air through their leaves, provide shade and ensure permeable substrate on the ground. Anyone entering a dense forest in midsummer can experience the climatic effect under the canopy up close. In order to create a similar, constantly moist and moderately warm climate during the growth phase of the fungal mycelium, a cover film is essential. In addition, the risk of mold growth is lower as long as the fungal culture is hermetically sealed.

The best way to grow mushrooms is on a dead tree trunk.

The tree trunk method has proven itself for growing mushrooms in the garden . While ready cultures and straw bales become unusable after a while due to rot, a well-placed stump provides the mushroom mycelium with "food" for survival for several years. In addition, the trunks in the garden look much better than a lump of straw.

And this is how it works: Take an approximately 50 centimeter long trunk or branch segment of a deciduous tree with a diameter of at least 15 cm and drill several holes all the way around. Sawing is also possible. The fresher the wood is, the better it is suitable for mushroom cultivation. The softer it is, the faster it goes but it doesn't last that long. It is important that it is still damp. If necessary, you can also put the trunk in fresh water for a few days. The holes should not go all the way through the trunk and should be about one and a half centimeters in diameter. A spiral-shaped distribution of the holes around the trunk has proven itself: a hole every 10 cm.The cavities are then filled with the fresh mushroom brood and sealed with a piece of cardboard and thumbtacks. The trunks can also be wrapped in foil. Then the mushroom cultivation is packed airtight in a plastic bag and stored in a shady place at temperatures of 15-25 degrees. If everything is done correctly, the mycelium should have completely overgrown the stem after three to four months.

If the first white mycelium tips are visible on the bark, the growth phase is over and the trunk is ready for the garden. A shady place with moist soil should be chosen for the placement. Here you dig the stumps vertically, about 15 centimeters should be enough so that they do not fall over. The trunk continues to draw the moisture necessary for mushroom growth from the moist soil, but you should still water regularly from above, especially on hot summer days. The advantage of the wood method is that the trunk only absorbs as much moisture as the mycelium needs to survive, the rest flows away. Waterlogging must be avoided, otherwise the mycelium will die.

This variant works best with mushroom brood from cereal grains. Mycelium patches on flat wooden disks work analogously . The discs do not last as long as whole trunks, but can be placed under bushes and small trees to save space. Mycelium sticks are preferred for straw bales.

A tree trunk completely interspersed with mycelium should form the first small fruiting bodies about two to six weeks after "unpacking". From this point on you can watch the mushrooms growing. Adult mushrooms can then be harvested in several harvests every 2-4 weeks. The trunks can stay outside during the winter; once overgrown, the mushroom farm is hardy. With proper care, a strain can deliver fresh mushrooms for several years before it is exhausted and decays.

In mushroom cultivation using the straw bale method , straw is watered for at least 48 hours and bound with a film to form a bale, in which mycelium sticks are then inserted. If you want to grow mushrooms in the basement all year round, this method is perhaps best served.

Here, after the growth phase, you can determine where the mushrooms should grow out through interfaces in the film.

STAGES OF MUSHROOM CULTIVATION

These steps are described in their naturally occurring sequence, emphasizing the salient features within each step. Compost provides nutrients needed for mushrooms to grow. Two types of material are generally used for mushroom compost, the most used and least expensive being wheat straw-bedded horse manure. Synthetic compost is usually made from hay and crushed corncobs, although the term often refers to any mushroom compost where the prime ingredient is not horse manure. Both types of compost require the addition of nitrogen supplements and a conditioning agent, gypsum.

The preparation of compost occurs in two steps referred to as Phase I and Phase II composting. The discussion of compost preparation and mushroom production begins with Phase I composting.

Phase I: Making Mushroom Compost

Phase I composting is initiated by mixing and wetting the ingredients as they are stacked in a rectangular pile with tight sides and a loose center. Normally, the bulk ingredients are put through a compost turner. Water is sprayed onto the horse manure or synthetic compost as these materials move through the turner. Nitrogen supplements and gypsum are spread over the top of the bulk ingredients and are thoroughly mixed by the turner. Once the pile is wetted and formed, aerobic fermentation (composting) commences as a result of the growth and reproduction of microorganisms, which occur naturally in the bulk ingredients. Heat, ammonia, and carbon dioxide are released as by-products during this process. The use of forced aeration, where the compost is placed on a concrete floor or in tunnels or bunkers and aerated by the forced passage of air via a plenum, nozzles or spigots located in the floor has become nearly universal in the mushroom industry.

Mushroom compost develops as the chemical nature of the raw ingredients is converted by the activity of microorganisms, heat, and some heat-releasing chemical reactions. These events result in a food source most suited for the growth of the mushroom to the exclusion of other fungi and bacteria. There must be adequate moisture, oxygen, nitrogen, and carbohydrates present throughout the process, or else the process will stop. This is why water and supplements are added periodically, and the compost pile is aerated as it moves through the turner.

The qualities of raw materials used to make mushroom compost are highly variable and are known to influence compost performance in terms of spawn run and mushroom yield. The geographical source of wheat straw, the variety (winter or spring) and the use of nitrogen fertilizer, plant growth regulators and fungicides may affect compost productivity. Wheat straw should be stored under cover to minimize growth of unwanted and potentially detrimental fungi and bacteria prior to its use to produce compost.

Gypsum is added to minimize the greasiness compost normally tends to have. Gypsum increases the flocculation of certain chemicals in the compost, and they adhere to straw or hay rather than filling the pores (holes) between the straws. A side benefit of this phenomenon is that air can permeate the pile more readily, and air is essential to the composting process. The exclusion of air results in an airless (anaerobic) environment in which deleterious chemical compounds are formed which detract from the selectivity of mushroom compost for growing mushrooms. Gypsum is added at the outset of composting at 40 lb per ton of dry ingredients.

Nitrogen supplements in general use today includes corn distiller's grain, seed meals of soybeans, peanuts, or cotton, and chicken manure, among others. The purpose of these supplements is to increase the nitrogen content to 1.5 percent for horse manure or 1.7 percent for synthetic, both computed on a dry weight basis. Synthetic compost requires the addition of ammonium nitrate or urea at the outset of composting to provide the compost microflora with a readily available form of nitrogen for their growth and reproduction.

The initial compost pile should be 5 to 6 feet wide, 5 to 6 feet high, and as long as necessary. A two-sided box can be used to form the pile (rick), although some turners are equipped with a "ricker" so a box isn't needed. The sides of the pile should be firm and dense, yet the center must remain loose throughout Phase I composting. As the straw or hay softens during composting, the materials become less rigid and compactions can easily occur. If the materials become too compact in the traditional Phase I process, air cannot move through the pile and an anaerobic environment will develop. The problem of an anaerobic center core in the compost has largely been overcome by using forced aeration.

Turning and watering are done at approximately 2-3 day intervals, but not unless the pile is hot (145° to 170°F). Turning provides the opportunity to water, aerate, and mix the ingredients, as well as to relocate the straw or hay from a cooler to a warmer area in the pile, outside versus inside. Supplements are also added when the compost is turned, but they should be added early in the composting process. The number of turnings and the time between turnings depends on the condition of the starting material and the time necessary for the compost to heat to temperatures above 145°F.

Water addition is critical since too much will exclude oxygen by occupying the pore space, and too little can limit the growth of bacteria and fungi. As a general rule, water is added up to the point of leaching when the pile is formed and at the time of first turning, and thereafter either none or only a little is added for the duration of composting. On the last turning before Phase II composting, water can be applied generously so that when the compost is tightly squeezed, water drips from it. There is a link between water, nutritive value, microbial activity, and temperature, and because it is a chain, when one condition is limiting for one factor, the whole chain will cease to function.

Phase I composting lasts from 6 to 14 days, depending on the nature of the material at the start and its characteristics at each turn. There is a strong ammonia odor associated with composting, which is usually complemented by a sweet, moldy smell. When compost temperatures are 155°F and higher, and ammonia is present, chemical changes occur which result in a food rather exclusively used by the mushrooms. As a by-product of the chemical changes, heat is released and the compost temperatures increase. Temperatures in the compost can reach 170° to 180°F during the second and third turnings when a desirable level of biological and chemical activity is occurring. At the end of Phase I the compost should: a) have a chocolate brown color; b) have soft, pliable straws, c) have moisture content of from 68 to 74 percent; and d) have a strong smell of ammonia. When the moisture, temperature, color, and odor described have been reached, Phase I composting is completed.

Phase II: Finishing the Compost

There are two major purposes to Phase II composting. Pasteurization is necessary to kill any insects, nematodes, pest fungi, or other pests that may be present in the compost. And second, it is necessary to condition the compost and remove the ammonia that formed during Phase I composting. Ammonia at the end of Phase II in a concentration higher than 0.07 percent is often inhibitory to mushroom spawn growth, thus it must be removed; generally, a person can smell ammonia when the concentration is above 0.10 percent.

Phase II takes place in one of three places, depending on the type of production system used. For the zoned system of growing, compost is packed into wooden trays, the trays are stacked six to eight high, and are moved into an environmentally controlled Phase II room. Thereafter, the trays are moved to special rooms, each designed to provide the optimum environment for each step of the mushroom growing process. With a bed or shelf system, the compost is placed directly in the beds, which are in the room used for all steps of the crop culture. The most recently introduced system, the bulk system, is one in which the compost is placed in an insulated tunnel with a perforated floor and computer-controlled aeration; this is a room specifically designed for Phase II composting.

The compost, whether placed in beds, trays, or bulk, should be filled uniformly in depth and density or compression. Compost density should allow for gas exchange, since ammonia and carbon dioxide will be replaced by outside air.

Phase II composting can be viewed as a controlled, temperature-dependent, ecological process using air to maintain the compost in a temperature range best suited for microorganisms to grow and reproduce. The growth of these thermophilic (heat-loving) organisms depends on the availability of usable carbohydrates and nitrogen, some of the nitrogen in the form of ammonia. These microorganisms produce nutrients or serve as nutrients in the compost on which the mushroom mycelium thrives and other organisms do not.

Completing Phase II in tunnels has become more popular in recent years. Tunnel composting has the advantage of treating more compost per ft2 compared to more expensive production rooms. When coupled with bulk spawn run, tunnel composting offers the advantage of more uniformity and greater use of mechanization. However, transfer of the finished compost from the pasteurization tunnel to the bulk spawn run tunnel may increase the risk of infestation of unwanted pathogens and pests compared to compost that remains in the same room. Thus, higher levels of sanitation may be required with tunnel composting compared to in-room composting.

It is important to remember the purposes of Phase II when trying to determine the proper procedure and sequence to follow. One purpose is to remove unwanted ammonia. To this end the temperature range from 125° to 130°F is most efficient since de-ammonifying organisms grow well in this temperature range. A second purpose of Phase II is to remove any pests present in the compost by use of a pasteurization sequence.

At the end of Phase II the compost temperature must be lowered to approximately 75° to 80°F before spawning (planting) can begin. The nitrogen content of the compost should be 2.0 to 2.4 percent, and the moisture content between 68 and 72 percent. Also, at the end of Phase II it is desirable to have 6 to 8 lb of dry compost per square foot of bed or tray surface to obtain profitable mushroom yields. It is important to have both the compost and the compost temperatures uniform during the Phase II process since it is desirable to have as homogenous a material as possible.

Spawning

As a mushroom matures, it produces millions of microscopic spores on mushroom gills lining the underside of a mushroom cap. These spores function roughly similar to the seeds of a higher plant. However, growers do not use mushroom spores to 'seed' mushroom compost because they germinate unpredictably and therefore, are not reliable. Fortunately, mycelium (thin, thread-like cells) can be propagated vegetatively from germinated spores, allowing spawn makers to multiply the culture for spawn production. Specialized facilities are required to propagate mycelium, so the mushroom mycelium remains pure. Mycelium propagated vegetatively on various grains or agars is known as spawn, and commercial mushroom farmers purchase spawn from companies specializing in its manufacture.

Spawn makers start the spawn-making process by sterilizing a mixture of millet grain plus water and chalk; rye, wheat, and other small grain may be substituted for millet. Sterilized horse manure formed into blocks was used as the growth medium for spawn up to about 1940, and this was called block or brick spawn, or manure spawn; such spawn is not used today. Once sterilized grain has a bit of mycelium added to it, the grain and mycelium is shaken 3 times at 4-day intervals over a 14-day period of active mycelial growth. Once the grain is colonized by the mycelium, the product is called spawn (Fig. 3). Spawn can be refrigerated for a few months, so spawn is made in advance of a armer's order for spawn.

Spawn is distributed on the compost and then thoroughly mixed into the compost. For years this was done by hand, broadcasting the spawn over the surface of the compost and ruffling it in with a small rake-like tool. In recent years, however, for the bed system, spawn is mixed into the compost by a special spawning machine that mixes the compost and spawn with tines or small finger-like devices. In a tray or batch system, spawn is mixed into the compost as it moves along a conveyer belt or while falling from a conveyor into a tray. The spawning rate is expressed as a unit or quart per so many square feet of bed surface; 1 unit per 5 ft2 is desirable. The rate is sometimes expressed on the basis of spawn weight versus dry compost weight; a 2 percent spawning rate is desirable.

Supplementation at spawning

In the early 1960s, yield increases were observed when compost was supplemented with protein and/or lipid rich materials at spawning, casing and later. Up to a 10% increase in yield was obtained when small amounts of protein supplements were added to the compost at spawning. Excessive heating and stimulation of competitor molds in the compost substantially limited the amount of supplement and corresponding benefit that could be achieved. It was these limitations that were overcome by the invention of delayed release supplements for mushroom culture (Carroll and Schisler 1976). The disadvantages associated with the supplementation of non-composted nutrients to mushroom compost at spawning were largely overcome by encapsulating micro-droplets of vegetable oil within a protein coat that was denatured with formaldehyde. Increases of as much as 60% were obtained. Today, several commercial supplements are available that can be used at spawning or at casing to stimulate mushroom yield.

Amendment of mushroom substrate with Micromax is another potential opportunity for growers to improve the yield capacity of their Phase II compost. Micromax contains a mixture of nine micronutrients including (percentage dry wt basis): Ca (12%), Mg (3%), S (12%), B (0.1%), Cu (1%), Fe (17%), Mn (2.5%), Mo (0.05%), Zn (1%), and inert ingredients (57.35%). Research has shown that approximately 70% of the yield increase observed is due to Mn. Commercial supplement makers have begun to add Mn to their delayed release nutrients for mushroom culture.

Once the spawn and supplement have been mixed throughout the compost and the compost worked so the surface is level, the compost temperature is maintained at 75-80°F and the relative humidity is kept high to minimize drying of the compost surface or the spawn. Under these conditions the spawn will grow - producing a thread-like network of mycelium throughout the compost. The mycelium grows in all directions from a spawn grain, and eventually the mycelium from the different spawn grains fuses together, making a spawned bed of compost one biological entity. The spawn appears as a white to blue-white mass throughout the compost after fusion has occurred. As the spawn grows it generates heat, and if the compost temperature increases to above 80° to 85°F, depending on the cultivar, the heat may kill or damage the mycelium and eliminate the possibility of maximum crop productivity and/or mushroom quality. At temperatures below 74°F, spawn growth is slowed and the time interval between spawning and harvesting is extended.

Phase III and Phase IV compost

Phase III compost is Phase II compost spawn run in bulk in a tunnel, and ready for casing when removed from the tunnel and delivered to the grower. If the Phase III compost then is cased and the spawn allowed to colonize the casing layer before sending to the growing unit or delivering to growers, it is called Phase IV compost. The successes of both Phase III and Phase IV compost depend, to a large extent, on the quality of Phase I and Phase II composts. Use of Phase III compost may also improve mushroom quality, as fragmentation of the colonized compost tends to improve initial color and mushroom shelf life. In recent years, the use of bulk Phase III compost has increased in popularity because it allows an increase in the number of crops a grower can expect from his production rooms. Phase II production on shelves allows an average of about 4.1 crops per year whereas growers using Phase III bulk spawn run compost averages about 7.1 crops per year. An additional gain can be made in the number of crops (10-12 crops per year) when

Phase IV is used (Dewhurst 2002; Lemmers 2003; Chang 2006).

MUSHROOM VARIETIES

In the United States, mushroom growers use three major mushroom cultivars: a) Smooth white hybrid - cap smooth, cap and stalk white; b) Off-white hybrid - cap scaly with stalk and cap white; and c) Brown - cap smooth, cap chocolate brown with a white stalk. Within each of the three major groups, there are various isolates, so a grower may have a choice of up to eight strains within each variety. Generally, white and off-white hybrid cultivars are used for processed foods like soups and sauces, but all isolates are good eating as fresh mushrooms. In recent years, the brown varieties have gained market share among consumers. The Crimini variety is similar in appearance to the white mushroom except it is brown and has a richer and earthier flavor. The Portobello variety is a large, open, brown-colored mushroom that can have caps up to 6 inches in diameter. The Portobello offers a rich flavor and meaty texture.

The time needed for spawn to colonize the compost depends on the spawning rate and its distribution, the compost moisture and temperature, compost supplementation, and the nature or quality of the compost. Complete spawn run usually requires 13 to 20 days. Once the compost is fully-grown with spawn, the next step in production is at hand.

Casing

Casing is a top-dressing applied to the spawn-run compost on which the mushrooms eventually form. A mixture of peat moss with ground limestone can be used as casing. Casing does not need nutrients since casing acts as a water reservoir and a place where rhizomorphs form. Rhizomorphs look like thick strings and form when the very fine mycelium fuses together. Mushroom initials, primordia, or pins form on the rhizomorphs, so without rhizomorphs there will be no mushrooms. Casing should be able to hold moisture since moisture is essential for the development of a firm mushroom. The most important functions of the casing layer are supplying water to the mycelium for growth and development, protecting the compost from drying, providing support for the developing mushrooms and resisting structural breakdown following repeated watering. Supplying as much water as possible to the casing as early as possible without leaching into the underlying compost provides the greatest yield potential.

Sphagnum peat moss is the most commonly used material for casing. Sphagnum can range from brown (young, less decomposed, loose textured, surface peat) to black (compact, more decomposed, deep dug) and may be processed differently at the harvest site. Milled peat is partially dried before packaging and transport while wet-dug peat is transported in a saturated form. Some growers prefer wet-dug peat because of the higher water holding capacity compared to milled peat.

Peat moss-based casing does not require pasteurization because the material is free from pathogens, weed molds and nematodes that may reduce mushroom yield. One 6-ft3 compressed bale when mixed with water and 40 lb of limestone will cover about 125 ft2 of compost surface at about 2 inches depth.

Casing inoculum (CI)

Casing inoculum is a sterilized mixture of peat, vermiculite and wheat bran that has been colonized by mushroom mycelium. It is mixed with casing to decrease cropping cycle time, improve uniformity of mushroom distribution over the bed and improve mushroom cleanliness. Mycelium from the CI colonizes the casing layer while it fuses with the underlying mycelium of the compost. This allows more breaks per crop or more crops per year.

Supplementation at casing

The addition of nutrients at casing was first tried in the early 1960s. Results showed that much greater amounts of nutrients could be added at casing than at spawning and that yield increases were almost proportional to the amount added. Although yield increases as high as 100% may be realized, certain potential problems and limitations exist for supplementation at casing. Weed molds, nematodes and pathogens must not be present in the compost when supplementing at casing. These organisms will be dispersed throughout the compost when it is fragmented prior to supplementation and can multiply very rapidly before the mushroom mycelium recovers its growth.

Managing the crop after casing requires that the compost temperature be kept at around 75°F for up to 5 days after casing, and the relative humidity should be high. Thereafter, the compost temperature should be lowered about 2°F each day until small mushroom initials (pins) have formed. Throughout the period following casing, water must be applied intermittently to raise the moisture level to field capacity before the mushroom pins form. Knowing when, how, and how much water to apply to casing is an "art form" which readily separates experienced growers from beginners.

Pinning

Mushroom initials develop after rhizomorphs have formed in the casing. The initials are extremely small but can be seen as outgrowths on a rhizomorph. Once an initial quadruples in size, the structure is a pin. Pins continue to expand and grow larger through the button stage, and ultimately a button enlarges to a mushroom (Fig. 5). Harvestable mushrooms appear 18 to 21 days after casing. Pins develop when the carbon dioxide content of room air is lowered to 0.08 percent or lower, depending on the cultivar, by introducing fresh air into the growing room. Outside air has a carbon dioxide content of about 0.04 percent.

The timing of fresh air introduction is very important and is something learned only through experience. Generally, it is best to ventilate as little as possible until the mycelium has begun to show at the surface of the casing, and to stop watering at the time when pin initials are forming. If the carbon dioxide is lowered too early by airing too soon, the mycelium will stop growing through the casing and mushroom initials form below the surface of the casing. As such mushrooms continue to grow, they push through the casing and are dirty at harvest time. Too little moisture can also result in mushrooms forming below the surface of the casing. Pinning affects both the potential yield and quality of a crop and is a significant step in the production cycle.

Cropping

The terms flush, break, or bloom are names given to the repeating 3- to 5-day harvest periods during the cropping cycle; these are followed by a few days when no mushrooms are available to harvest. This cycle repeats itself in a rhythmic fashion, and harvesting can go on as long as mushrooms continue to mature. Most mushroom farmers harvest for 35 to 42 days, although some harvest a crop for 60 days, and harvest can go on for as long as 150 days.

Air temperature during cropping should be held between 57° to 62°F for good results. This temperature range not only favors mushroom growth, but cooler temperatures can lengthen the life cycles of both disease pathogens and insects pests. It may seem odd that there are pests that can damage mushrooms, but no crop is grown that does not have to compete with other organisms. Mushroom pests can cause total crop failures, and often the deciding factor on how long to harvest a crop is based on the level of pest infestation. These pathogens and insects can be controlled by cultural practices coupled with the use of pesticides, but it is most desirable to exclude these organisms from the growing rooms.

The relative humidity in the growing rooms should be high enough to minimize the drying of casing but not so high as to cause the cap surfaces of developing mushrooms to be clammy or sticky. Water is applied to the casing so water stress does not hinder the developing mushrooms; in commercial practice this means watering 2 to 3 times each week. Each watering may consist of more or fewer gallons, depending on the dryness of the casing, the cultivar being grown, and the stage of development of the pins, buttons, or mushrooms. Most first-time growers apply too much water and the surface of the casing seals; this is seen as a loss of texture at the surface of the casing. Sealed casing prevents the exchange of gases essential for mushroom pin formation. One can estimate how much water to add after first break has been harvested by realizing that 90 percent of the mushroom is water and a gallon of water weighs 8.3 lb. If 100 lb of mushrooms were harvested, 90 lb of water (11 gal.) were removed from the casing; and this is what must be replaced before second break mushrooms develop.

Outside air is used to control both the air and compost temperatures during the harvest period. Outside air also displaces the carbon dioxide given off by the growing mycelium. The more mycelial growth, the more carbon dioxide produced, and since more growth occurs early in the crop, more fresh air is needed during the first two breaks. The amount of fresh air also depends on the growing mushrooms, the area of the producing surface, the amount of compost in the growing room, and the condition or composition of the fresh air being introduced. Experience seems to be the best guide regarding the volume of air required, but there is a rule of thumb: 0.3ft/ft2/hr when the compost is 8 inches deep, and of this volume 50 to 100 percent must be outside air.

Ventilation is essential for mushroom growing, and it is also necessary to control humidity and temperature. Moisture can be added to the air by a cold mist or by live steam, or simply by wetting the walls and floors. Moisture can be removed from the growing room by:

1) Admitting a greater volume of outside air;

2) Introducing drier air;

3) Moving the same amount of outside air and heating it to a higher temperature since warmer air holds more moisture and thus lowers the relative humidity. Temperature control in a mushroom growing room is no different from temperature control in your home. Heat can originate from hot water circulated through pipes mounted on the walls. Hot, forced air can be blown through a ventilation duct, which is rather common at more recently built mushroom farms. There are a few mushroom farms located in limestone caves where the rock acts as both a heating and cooling surface depending on the time of year. Caves of any sort are not necessarily suited for mushroom growing, and abandoned coalmines have too many intrinsic problems to be considered as viable sites for a mushroom farm. Even limestone caves require extensive renovation and improvement before

they are suitable for mushroom growing, and only the growing occurs in the cave with composting taking place above ground on a wharf.

Mushrooms are harvested in a 7- to 10-day cycle, but this may be longer or shorter depending on the temperature, humidity, cultivar, and the stage when they are picked (Fig. 6). When mature mushrooms are picked, an inhibitor to mushroom development is removed and the next flush moves toward maturity. Mushrooms are normally picked at a time when the veil is not too far extended. Consumers in North America want closed, tight, and white or brown (Crimini) mushrooms while open browns (Portobello) are preferred by some consumers. The maturity of a mushroom is assessed by how far the veil is stretched, and not by how large the mushroom is. Consequently, mature mushrooms are both large and small, although farmers and consumers alike prefer medium- to large-size mushrooms.

Picking and packaging methods often vary from farm to farm. Freshly harvested mushrooms must be kept refrigerated at 35° to 45°F. To prolong the shelf life of mushrooms, it is important that mushrooms "breathe" after harvest, so storage in a non-waxed paper bag is preferred to a plastic bag.

A question frequently arises concerning the need for illumination while the mushrooms grow. Mushrooms do not require light to grow; only green plants require light for photosynthesis. However, growing rooms can be illuminated to facilitate harvesting or cropping practices.

Nutrients

Mushrooms are a good source of numerous nutrients. They are an excellent source (contain over 20% of the RDA in a serving) of selenium, riboflavin (vitamin B2) and copper and are a good source (contain over 10% of RDA) for niacin (vitamin B3), pantothenic acid (vitamin B5) and potassium. Criminis also contain rich amounts of thiamin (Vitamin B1), zinc, vitamin B6, protein, folic acid, fiber, manganese and magnesium. On the other hand, mushrooms are low in fat, sodium and calories.

Reference serving size of 84 g for raw mushrooms. Mushrooms are not a significant source of saturated fat, Trans fat, cholesterol, sugars, vitamin A and calcium. Source: Mushroom Council.

Vitamin D

Recent research has shown that when UV light is shined on mushrooms, there is a major boost in the vitamin D2 content of the mushrooms. A single serving of mushrooms will contain over 800% of the recommended daily allowance (RDA) of vitamin D2 once exposed to just five minutes of UV light after being harvested. This may be a convenient way for people who do not eat fish or drink milk to obtain their daily requirement of vitamin D.

Dietary fiber (DF)

Mushrooms contain numerous complex carbohydrates including polysaccharides such as glucans and glycogen, monosaccharides, disaccharides, sugar alcohols and chitin. Most polysaccharides are structural components of the cell walls (chitin and glucans) and are indigestible by humans; thus they may be considered as dietary fiber. Dietary fiber may help to prevent many diseases prevalent in affluent societies. Portobello mushrooms contain a higher level of DF than the white variety of mushrooms.

Selenium

A serving (3 ounces) of Crimini mushrooms provides almost one-third of the RDA for selenium, according to the USDA National Nutrient Database. Selenium has been shown to decrease prostate cancer by more than 60% according to findings from the Baltimore Longitudinal Study on Aging. Men with the lowest blood selenium levels were 4-5 times more likely to have prostate cancer than those with the highest selenium levels and that selenium levels tend to decrease with age.

Selenium levels can be reliably increased in mushrooms by adding sodium selenite to mushroom compost. Some commercial supplement makers are now adding this compound to their delayed release nutrients for mushroom culture.

Potassium

Crimini mushrooms are a good source of potassium, an element that is important in the regulation of blood pressure, maintenance of water in fat and muscle, and to ensure the proper functioning of cells. A 3-ounce Portobello contains more potassium than a banana or an orange. To date, attempts to enhance the potassium content of mushrooms have met with only limited success.

Antioxidants

Portobello and Crimini mushrooms are good sources of antioxidants and rank with carrots, green beans, red peppers and broccoli as good sources of dietary antioxidants. They are rich sources of polyphenols that are the primary antioxidants in vegetables and are the best source of L-ergothioneine (ERGO) - a potent antioxidant only produced in nature by fungi. Crimini mushrooms contain over 15 times more ERGO than the previously best-known dietary sources of ERGO.

Environmental Concerns

Odors

Nuisance complaints, a result of mushroom compost preparation in close proximity to residential areas, are a problem for some mushroom farms. Offensive odors associated with the preparation of mushroom compost are the primary reasons for these complaints. A combination of suburbanization and the heightened sensitivity of the general population to environmental issues have focused public attention on this issue. Growers have adopted several measures to reduce the environmental impact of mushroom farming, including the practice of forced aeration of Phase I compost contained in bunkers or tunnels. However, the issue of offensive odor generation continues to place pressure on mushroom growers.

Disposal of post-crop mushroom compost

After the last flush of mushrooms has been picked, the growing room should be closed off and the room pasteurized with steam. This final pasteurization is designed to destroy any pests that may be present in the crop or the woodwork in the growing room, thus minimizing the likelihood of infesting the next crop.

Post-crop mushroom compost (MC) is the material left over after the crop has been terminated (Fig. 8). It has many uses and is a valued product in the horticultural industry. One of the major uses of MC is for suppression of artillery fungi in landscape mulch. The artillery fungi grow rapidly throughout moist landscape mulch, and produce sticky spore masses about the size of a pinhead. These spores are forcibly discharged toward light colored surfaces such as house siding and cars. Once the spores dry they are nearly impossible to remove without leaving an unsightly brown stain on the surface. The incorporation of 20-

40% MC into mulch effectively suppress the artillery fungi.

MOST POPULAR MUSHROOM TO GROW

There are many wonderful different types of mushrooms to choose from, but don't make this decision lightly- there are too many things to consider.

First, you have to consider of you want to grow just one type of mushroom, or if you want to grow multiple species. There are advantages and disadvantages to both.

Growing just one type of mushroom means that you have fewer complications to deal with when creating grain spawn and trying to time the mushroom cycle. It also allows you to dial in the environmental conditions in your grow room (if you only have one) without compromising. That being said, if you are trying to sell mushrooms, it might be beneficial to have some variety.

You also have to consider that different mushrooms will have different shelf lives, handling ability and utility in the kitchen. They will also have different market acceptability and demand a unique price per pound, depending on where you intend to sell your mushrooms.

Basically, it depends. And it's up to you. But I would like to offer some suggestions.

Oyster Mushrooms

These are probably the most widely grown mushrooms by beginners. Oyster mushrooms are consumed less than button mushrooms in the West, but they're very popular in Asian countries for everyday cooking.

They have an appearance that you may not have seen before in a mushroom. That's because in nature they grow on the side of trees, so they have a large flat cap with little or no stem.

Blue Oyster Mushrooms

Blue oyster mushrooms are easy to grow, especially when growing on straw logs where they can rapidly produce big yields. Oyster mushrooms are also relatively well known and available, although you will rarely find fresh, good looking oyster mushrooms at most grocery stores. This means that you might be able to demand a good price for your mushrooms. Pleurotus species mushrooms do not take well to shipping long distances, so you might have an edge over bigger commercial growers. Check out your local scene and see if the price they demand will be worth your effort.

King Oyster Mushrooms

King Oyster mushrooms are also quite easy to grow, although they tend to grow better on supplemented sawdust than they do on straw. The best part about the king oyster mushroom is their long shelf life. Typically, a properly harvested king oyster mushroom will be sellable for up to two weeks, which is not common for most other gourmet species. Unfortunately, this also means that king oyster mushrooms are easily shipped long distances. Chances are that you will be competing with cheap commercially grown king oyster mushrooms from Chinese or Korean suppliers. However, these kings are generally grown with small caps and big fat stems. You could differentiate yourself by growing kings in a bright, low CO_2 environment to produce a unique looking king with big caps and small stems.

Lions Mane Mushrooms

Consistently growing fresh Lions Mane Mushrooms might really set you apart from the crowd, as these mushrooms are not often sold to consumers from commercial growers. This is likely because there delicate teeth require careful handling and packaging. However, a small local grower might be able carefully harvest this mushroom and deliver it straight to the customer. Although these mushrooms are unknown to a lot of chefs and consumers, they are generally well liked when experienced and could provide a unique market opportunity.

Button / Crimini / Portobello Mushrooms

If you've only ever eaten one type of mushroom in your life, I can almost guarantee that it's this one. These mushrooms are actually all the same species. The only difference is how long they're allowed to grow before being harvested.

When these mushrooms first appear from the mycelium, they start off as button mushrooms. As they get a bit larger, they develop a brown color and become what is usually categorized as crimini in shops.

Finally they fully grow into a portobello mushroom. These are the big brown mushrooms with dark gills underneath that are often sliced up or grilled whole. They have a tougher, more meaty texture.

Shiitake Mushrooms

Shiitake mushrooms have a smoky, earthy flavor and a texture similar to portobellos. In addition to being delicious, they also offer several health benefits including compounds that can help to lower cholesterol.

In stores, shiitake are normally sold dried. But eating them fresh is a real treat. Shiitakes are often grown outdoors on logs.

Enoki

Enoki mushrooms are very small with long stems. They grow together in tight clumps. If it weren't for their tiny caps, they would look almost like spaghetti! Enoki are very compact so you can grow them without much space required. They're typically grown in jars.

Maitake

Maitake mushrooms are another variety that is both delicious and also has strong nutritional and health benefits. Don't be confused by this mushroom's nickname "hen of the woods." The name comes from its appearance which people say resembles the ruffled feathers of a hen, and doesn't refer to its taste.

Maiitake has a very strong, earthy taste and I'd recommend trying some to make sure you like it before you spend time growing it!

HARVESTING AND STORAGE OF MUSHROOM

No matter what mushroom you decide to grow, you are going to need to properly harvest and store before delivering it to your kitchen or your customers. Harvesting your mushrooms is rewarding, but it is also a pretty time consuming process which takes a while to start doing efficiently.

You will want to treat harvesting your mushrooms like any other aspect of food preparation. This means only using clean tools and equipment, wearing clean clothes and gloves, and having a wash sink nearby. Ensure that your knife is cleaned often, as it doesn't take long for harmful bacteria to grow on your knife, which could get passed on to your final product. You also might want to consider wearing a really good mask, especially when harvesting Pleurotus species oyster mushrooms.

It is common for mushroom growers to develop an allergy to the spores of these mushrooms, which can get worse over time.

Some growers choose not to grow pleurotus for this reason. Wearing a good mask when in the grow room is your best defense against developing this allergy.

Harvested king oyster mushrooms

Generally, harvesting involves cutting the mushroom fruit body off at the stem and removing any remaining substrate material. This is where you can really shine above your competition, by making sure that you are only sending your customers cleanly harvested, high quality mushrooms. Generally, you will want to have a scale nearby when harvesting so that you can weigh your mushrooms to get an idea of your yield and how much you are actually producing.

Often, substrate blocks and straw logs can be harvested multiple times, with diminishing returns and higher chances of contamination after the 2nd or 3rd harvest. When the substrate is finally spent, you will need to properly dispose of it in a compost pile or similar.

This is a huge consideration that is often overlooked by new growers, as the amount of spent substrate can become a large pile (and large problem!) rather quickly. You might need to work something out with a local land owner in order to properly dispose of your substrate.

MUSHROOM PEST AND DISEASE

The first report of Indian fungal nematode damage comes from Himachal Pradesh. An intensive study of mushroom growing in Solan showed 84.4% of nematode infections, resulting in extremely poor blushing in the early stages, followed by complete failure. The fungal worm problems are unique in that not only do nematodes fully meet the ecological requirements of culture, but they multiply very quickly and cause up to 100% plant loss. Therefore, there is a growing need for scientifically and economically viable mushroom-growing technologies.

Mushroom-related nematodes fall into five categories based on eating habits. Among these, myceliophagic and saprophagic nematodes are economically important, as common fungi such as Agaricus spp. They are very pathogenic. Mycetophagous women and men, as well as infectious women of the Iotonchium class, not yet known in Japan, were found by the pearl shellfish, Pleurotus ostreatus.

H. atrocaerulea fungi were first cultivated on rice straw in order to ensure the long-term supply of basidiocarps and to test their nematicidal potential. In vitro tests with crude extracts of basidiocarps on Meloidogyne spp. 48% mortality of infectious larvae of the plant-parasitic nematode was found.

Iotonchium ungulatum is a causative agent of Pillurotus ostreatus oyster mushroom Gillknot disease. Females of this insect parasite parasite are described. The insect-parasitic female nematodes settle and lay eggs in the globe of the fungus mosquito Rhymosia domestica.

Fungus-feeding nematodes: 21 species of nematodes have been reported that damage fungi in two parts of the world, Aphelenchida and Tylenchida. Of these 20, four genera belong to the Aphelenchoides, Aphelenchus, Paraphelenchus and Seinura groups under the Aphelenchida order, and a kind of Ditylenchus myceliophagus to the Tylenchida order. In India, eight species of Aphelenchoides and Ditylenchus myceliophagus have been discovered in fungal beds.

Aphelenchoides spp:

I. Aphelenchoides agarici: It is derived from Himachal Pradesh, and the most preferred host is A. bisporus. Failing this, the nematode may multiply on other fungi of compost media, such as Trichoderma, Trichothecium, and Gillaminello. It has a short life span of 8 days, and many generations repeat over the course of the season.

II. Aphelenchoides asterocaudatus: This was first reported by Bahl and Prasad from the harvest beds in India.

III. Aphelenchoides composticola: Almost all mushroom producing countries in India are affected by this nematode. It is the most common nematode in almost every mushroom-producing country in the world. Three mycophagous nematodes, i. H. Aphelenchoides composticola, A. avenae, and Ditylenchus myceliophagus, as well as some saprophages (Rhabditis sp.) And carnivorous (Seinura sp.) Nematodes were found in compost samples in the village of Sonipat, Haryana. A. composticola was the dominant species with a prevalence of 50% and D. myceliophagus with 22.7%. Analysis of soil samples from fungal compost revealed Aphelenchoides composticola (10 persons / 100 ml) in only one case. However, mushroom production was highest when 1% neem seed extract and 400 ppm achook were combined during spawning. The population of Aphelenchoides composticola nematodes also decreased drastically in treated boxes compared to untreated controls. 20 grams of Neem cake was most effective in reducing the population of Aphelenchoides composticola, while 10 grams and 16 grams of Neem leaves had a manual effect on Agaricus bisporus in terms of yield potential, but Achook was found to be fungal toxic at higher doses, for example, it was obvious the number of fruiting bodies and the yields. It reproduces very quickly, with a short life cycle of 8 days at 23 ° C, ten days at 18 ° C and 18 days at 13 ° C.

IV. ARC. Aphelenchoides minor: In Srinagar (Jammu and Kashmir), India, only a few of these types of yarn spheres have been discovered from mushroom picking points in mushroom beds.

V. Aphelenchoides myceliophagus: In India, this nematode was first reported as a fungal cultivation in Solan (Himachal Pradesh). It is a very destructive and pathogenic species such as A. composticola.

VI. Aphelenchoides neocomposticola: The nematode was first found in an Indian Shimla mushroom farm when it damaged the mycelium. Less destructive than A. agarici, A. composticola and A. myceliophagus.

VII. Aphelenchoides sacchari: We have found that a highly pathogenic nematode reduces the yield of sporophore by 94.5% and delays the formation of fruit under natural conditions by about two weeks. This species was first discovered in India from white mushroom harvest beds. It is a bisexual species with a life cycle of 12 days. The presence of fungal nematodes in the growing system posed a serious threat and the most difficult problem to solve during cultivation, as they were the only ectoparasites in the realm of nematodes that caused a frequent and complete failure of the fungi. Aphelenchoides agarici and Ditylenchus myceliophagus, also known as myceliophagic nematodes, caused maximum damage at various stages of culture. Yield loss was 100% in A. sacchari and 70% in D. myceliophagus. Some herbal ingredients, such as Neem Azadirachta Indica, coconut, castor bean, and peanut cakes, are quite effective and have no adverse effects on culture [8]. It multiplies rapidly from egg-to-egg-to-egg generation in just 12 days. Females have a fertility period of 28-30 days.

VIII. Aphelenchoides swarupi: This mushroom-consuming species was first found in the Ambala (Haryana) mushroom farm in India, where it was initially observed with wet, wet tap heads, but no fruit products. The beds average approx. 10,000 people / 100g of compost were counted in bed, indicating a total crop failure.

IX. Aphelenchus avenae: This nematode is often found in small quantities in soil compost and envelope samples from cultivated beds. Three mycophagous nematodes, namely Aphelenchoides composticola, Aphelenchus avenae and Ditylenchus myceliophagus, and some saprophagous (Rhabditis sp.) And predatory (Seinura sp.) Nematodes are found in the fungal fungus (Agaricus spp.) And Oyster. dominant species with an 80% frequency, followed by Aphelenchus avenae and D. myceliophagus and rhabditis sp., with a prevalence of 67, 13, and 13 in registered shellfish fungus.

Seinura winchesi: In India, Seinura sp. was found on a mushroom farm harvest bed in a solan (H.P., India). In this case, sporophore yields decreased, although initial mycelial growth was normal.

Tylenchids: D. myceliophagus is a common nematode found in almost every fungus producing belt in the world. In India, it has been found to be common in mushroom farms in Jammu and Kashmir, Punjabi, Karnataka, Andhra Pradesh, Himachal Pradesh and Tamil Nadu. Initially identified as D. a destroyer, he was later called Goodey D. myceliophagous. It is one of the most deadly species that can induce 15 to 70% mycelium depletion within 60 to 15 days after initial inoculation of 10 to 1,000 individuals. This species is able to reproduce as rapidly as A. sacchari and is less pathogenic than A. composticola and A. sacchari. Ditylenchus myceliophagus, Aphelenchoides composticola, Rhabditidae, Mesorhabditis sp., Pellioditis sp. and Prodontorhabditis sp. is registered in mushroom farming, from Wenzhou, Zhejiang, China. Samples of compost, cover and soil from 22 mushroom farms in 15 villages in the Sonipat district of Haryana

(India) mycophagous nematodes consist of Aphelenchoides composticola, Aphelenchus avenae, and Ditylenchus myceliophagus

The nematic activity of the 2-hydroxy-1-naphthalenyls, the imines derived from the 2-hydroxynaphthalindehydes, and the anilines and their derivatives has been tested against fungal anthrax fungi. Manrao and Kaul studied the 15 compounds tested. ppm was used. Further studies on nematicide concentration and exposure time showed that two of the compounds tested would be suitable for further study.

Soapworms: Rhabditida sapworms are found in almost all fungal farms and heavily damaged fungal beds in India. The most important genera are rhabditis, caenorhabditis, cephalobus, panagrolaimus, diplogaster and acrobeloides. Karanj cake effectively reduced the population of A. composticola by 2 and 4% and coconut, Niger and Neem cake by 4%. Coconuts, Karanj and Neem were as effective at 2 and 4% and sesame at 4% as chemical treatments for rhabditis spp. For the general public. The greatest development (1500 / g) of rhabditis cucumeris was observed when nematodes were less introduced (900 / g) before spawning, when infection and spawning occurred simultaneously.

Predatory nematodes: These nematodes, usually mononchids, are often found in fungal beds, but their distribution depends primarily on the population density of their prey. They play an important role in mushroom production as they feed on other nematodes, including aphelenchides.

Mushroom Mosquitoes (Sciarids): Adult Sciarids are small (3-4 mm), fine, double winged flies, dark gray / black, with large eyes and long filiform antennas. Women are generally taller than men and often lift their bellies with eggs. Adults do not fly easily, but jump quickly over a growing surface with a short jump. You can also "stay" indoors and on the walls of production houses.

Adult flies are found throughout the year in mushroom-producing facilities, but are most common from mid-January to mid-March. A mated female can average 150 to 170 white oval eggs, individually or in groups, within the growing substrate. The incubation is three to seven days depending on temperature.

Sciarid larvae are white, elongated, feet without legs, with striking, shiny black heads. At this stage, the larvae feed on the development of the mycelium and dig uncontrollably into the needles and small buttons that form spongy material. The mature larvae are approx. They are 8.0 mm long and can remove mycelial adhesions at the base of the stem. The development of the larvae depends on the temperature, but it is approx. After 12-15 days, the larvae turn into pupae. This is an inactive phase that is not feeding. Growth usually occurs 5-7 days after sunbathing.

In a 125g envelope, a sciarid larvae caused an average 0.5% yield loss. Since the cost of recommended measures to control sciarids is 0.5% of the value of the harvest, this is an economic threshold for this pest: Fungal mosquito problems can occur due to excessively wet conditions and sick roots should alert growers to a bad culture.

Phorids: Adult Phorids are slightly smaller (2-3 mm) but more robust than Sciarids. Darker in color, humped in appearance, no difference between male and female flies. Adult flies are usually located on the surface of the compost or in the immediate vicinity of the growing field. They are very active in the presence of light, with characteristic fast jerking motion. The flight of adult forums is temperature dependent and they cannot fly when the air temperature is below 12 ° C. Therefore, wild populations generally do not attack mushroom production between November and February. Each female can lay 50 eggs in the immediate vicinity of the development of mycelia. The larvae are cream-colored, feet without legs, without a distinct capsule. The front area narrows to one point, while the back area is blunt with small protrusions. The duration of fatty development depends on temperature and can vary from 15 days (24-27 ° C) to 50 days (16-21 ° C). The development of the larva is around the time of development. 1/3, the rest is spent on babies.

Mosquitoes (Cecids): There may be two types of Cecid flies, which usually occur in fungal houses: Heteropeza (white) and Mycophila (orange). Adult flies are rarely seen because they are too small, but pedogenic larvae are seen in large numbers. Farmers' knowledge and practice of industrial hygiene reduces their contamination to a negligible level. Increased use of deep pits also helps because flies are unable to lay eggs. When entering the courtyard they have a great potential for corruption because they are very stubborn and very difficult to destroy.

The larva is usually white, oblong, smooth and spindle-shaped, 1.5 to 2.35 mm long, with two blemishes, showing an "X" appearance. Young larvae feed on growing mycelium and open a hyphae bundle.

Behavior of mushroom flies:

• After mating, adult females are attracted to the development of the fungal mycelium in the broom compost.

• Compost remains attractive to adult females during the breeding season and is particularly sensitive during the second week.

• The development of mycelium after the vagina will once again appeal to adult women.

• All females may lay 50 eggs in the immediate vicinity of the development of mycelia.

• Life cycle time depends on temperature and environmental conditions associated with harvest times.

• The combination of elevated air temperature, spawning and case study periods facilitates the completion of a life cycle within 24 to 26 days.

• Lower temperatures after fracture and at harvest time can extend the life cycle by 40-50 days.

Mites: Usually mites occur in straw and manure and are transferred to mushroom houses. Most species are useful in mushroom production because they feed on nematodes (nematodes) and other mites. You can live on other fungi (weeds and indicator shapes) that are found in mushroom cultivation, though some can cause damage. Mites can feed on fungal mycelium and fungi, where they can cause surface discoloration.

Tarsonemid mites:

These mites are light brown and so tiny that they can only be seen under a microscope. When fed completely, they cause damage

Based on the species of fungus, the breeder will find out if these mites are present, since the stalk of the fungus has a reddish-brown color. In the event of a serious infection, the entire fungal base may become detached from the growth area.

The mites usually get into the compost during the passage phase in order to adhere to the sciarid flies. These migration stages usually occur when mites are overcrowded.

If mites are present in the cultivation house, little can be done. Therefore, effective composting and peak heating are required to ensure that they are destroyed during the pasteurization process. Good hygiene should be maintained near mushroom production, especially when plant debris is removed and there is no stagnant water.

Red-bellied mites / Dwarf mites:

These mites are often associated with the Penicillium and Trichoderma molds that feed on them. They are not considered as primary pests; Their presence usually indicates that trichoderma (green mold) is present in the compost. This indicates that the compost is unsatisfactory. They do not feed on fungi. These mites may be able to develop an intermediate region, called hypopus, in which they flatten and develop a suction plate to adhere to moving objects such as flies. Mites are angry with fungi at this point. The mites are yellowish-brown, 0.25 mm long, with a smooth appearance; They are also capable of rapid reproductive rate. As mentioned above, these mites are secondary pests and often swarm in the shell and on the surface of the fungus. Where this occurs, the presence of the fungi makes them unsaleable. These mites can also spread Trichoderma spores from bag to bag.

Tail tail: A water lily is a group of very small and tiny (1-2 mm) wingless insects. They often occur in enormous numbers on the surface of the water, in snow, in mushroom sheds, in or near unattended objects in the garden, and in other wet areas. Occasionally they break into the houses and are especially found in cellars, bathrooms, and kitchens. Wet ambient conditions are preferred because the spring ears breathe through the cuticle. If your habitat gets too dry, they actively seek a better environment. You can move it into the mushroom house through indoor window grilles, open doors, vents or on the floor. They can also occur in cracks and crevices or in cavities with wet walls. Springtails feed on algae, fungi, and rotting plants.

Spring spikes cause damage to mycelium and button sporophores, as well as oyster mushrooms. Adults have a basic color with purple stripes on the side of the body. Dark and rounded scales are present throughout the body. Dark and rounded scales are present throughout the body. Female eggs are laid individually or on a stack of wet rice straw, compost, and mushrooms. Eggs hatch in 30 days at 30 ° C. An adult has a lifespan of 70-80 days at 26 ° C and is active throughout the year.

In the mushroom house, these insects either feed on the mycelium in the compost or attack the fruit bodies. The buttons on the feeding area show slight dots and blushing. It also feeds on gills, thus observing the destruction of the gill lining.

Beetles: Most Staphylinidae is an adult or larval predators or optional predators. However, minorities feed only on plant materials, including fungi. They observed how the oyster mushroom was injured. Adults are attracted to the smell of dead mushrooms and the laying of eggs in discarded or overgrown mushrooms. Flies feed on soft gills and complete their life cycle within three weeks. Removing waste and debris from the mushroom house and surrounding areas prevents adults from laying eggs. The population is being monitored. Ripe / sown fungi should not be left in bed.

Preventive measures: Given that fungal cultivation limits pest control, the introduction of preventive measures may be a better way to prevent pests of insects and nematodes.

Fungal diseases Mild mold or spider web: - Dactylium dendroides Symptoms: - The envelope surface has a loose, white, spider web mold. At first, it turns white and later turns pink with age. The main sources of

infection are soil, air, wet surfaces, and high humidity.

Controls: - Good ventilation and avoid excessive humidity. PCNB (Pentachloro-nitrobenzene) is 0.1% and Dithan Z-78 is 0.2%.

Brown gypsum form Papulospora byssina Symptoms: - Cultivated trees are

first cloudy white and later brown. Pilz originally grows in compost. The compost is too wet at high temperatures. (28-32) c promotes infection by spawning and pruning above 18 ° C.

Check: - Keep the spawn running at a suitable temperature and cut off 2% formalin

White gypsum tip: - Scopulariopsis fumicola Symptoms: - Initially very similar to the eyebrow mold, but later becomes pink. Too much water at the anaerobic temperature of composting causes fungi to grow

Verification: - Formalin 2% and Dithan Z-78 0.2%

Olive species Chaetomium olivacearum Symptoms: - Appear in compost or spawn before shelling. It changes from white to olive green. Improper pasteurisation and inadequate ventilation lead to this.

Checking: - The temperature is kept below 60 ° C during pasteurization. Spray 0.2% thiram and 0.05% captan onto the trays.

Ink cap: - Coprinus lagopus and C. comatus Symptoms: - Appearance of a long, cylindrical stem with a small, thin cap that turns into black ink. Presence Indicates the presence of ammonia in the compost.

Verification: - Compost bins must be decontaminated — Re-Pasteurization of spawning trays at 60 ° C for 2 hours, as well as spawning and recycling.

Green mold Symptoms of Trichoderma viride: - Manifests itself as green patches on spawning and skin. Controls pencil formation and thus reduces yield. The fungus grows on dead material and dead fungal tissue. Improper pasteurization and high humidity are also responsible.

Inspection: - Spray 0.05% benol

Tropical Fungal Disease: - Pseudobalsamia microspora

Symptoms: - Round, cream-colored and wrinkled, convex surface with a good appearance. As it matures, it turns reddish-brown and releases spores. Lack of ventilation and high humidity lead to this.

Check: - Spawning and slaughter bed temperatures <18 ° C. Avoid high humidity

Bladder disease: - Mycogone perniciosa Symptoms: - Dense white mycelial carpet that reduces yield. Swallon shank and the smallest cap at an early stage. Unpasteurized compost also leads to this.

Verification: - Sterilized beds with 2% formalin. Spray Dithan Z-78 at 0.2% and Benlat at 0.05%.

Dry bladder, brown spot disease: - Verticillium malthousei or V. psialliste

Symptoms: - Causes a brown spot on the cap, which leads to irregular spots. In severe infections, the fungus may be distorted. The affected fungi become skin-like

Check: - Dithan Z-78 @ 0.2% 3 times on the housing. Check the housing's high temperature and proper ventilation.

Bacterial disease

Bacterial stain; Pseudomonas tolaassi

Symptoms: - The mushroom cap has brown, slightly sunken patches, irregular, yellow to dark brown. The main source of infection is infectious soil, which is separated by flies, mites, and nematodes.

Checking: - Stop soil sterilization and proper ventilation. Use of chlorinated water. Spray terramycin at a concentration of 9 mg / f on the beds.

Wet stain / acid rot. Bacillus spp. Heat-resistant endospores. Slightly gray or mucus-brown brown mucus, soaked at room temperature 12 to 24 hours before sterilization

Viral disease: Various diseases, such as La France (Sindenhauser, 1950) Brown disease and water streaks (Gandy, 1960) Disease X (knee bones et al., 1962) Fatal disease (Gandy and Hollings, 1962) Symptoms of viral disease: fungi shrunken, skinny and brown in color. The tapes become wet and gray under wet conditions. Delayed needle formation Creams and cream-colored mushrooms suffer less from the pure white variety.

Insect pests: - Fungi are attacked by insects and pests. Spring tails, stem flies, scary flies, mites, nematodes. They lay eggs and larvae in compost, feed on mycelium and dig into the stalk of the fungus

Sciarids: - Lycorella fenestralis Scaria carpophilla S. agaria Dark, cylindrical body with long antennae. Larvae are more harmful than adult animals and feed on compost. The larvae create a tunnel-like cave in the fungus.

Control: - Strict hygiene in the mushroom house. Correct twists during the composting process ation Impregnation of malathion ion 0.01%, chlorophenvinphos 208 ml / t

Spring tail Megaselia nigra M. Agrici M. bovistra Not visible to the naked eye. In bulk, they look like gun powder. Most often, they eat mycelium and attack the stalk and caps.

Inspection: Clean cultivation, proper pasteurisation of compost, and coverings. Use of 0.05% malathion as a spray for disinfection. Spray dichlorvos with 0.025 to 0.05% concentrate. in the spawning stage

Phorids: These severely damage the mycelium and sometimes lead to tunnels. The front is tapered. Lava causes far more damage than adults.

Control: - Strict hygiene in the mushroom house. Correct twists during composting. Mix with compost at spawning diazinon @ g / ton. Formalin is 0.2% of the case

Mites: - Mycophage mites: - Red pepper acne, usually eating weeds. Saprophage mites: These are translucent, long, long hairs in the body that lean on the mycelium and cause shrinking caps. Causes complete destruction of button buttons and tropical fungi. Introduced into the compost of flies.

Mycophage mites: - Saprophage mites: -

Inspection: - Cleaning the mushroom house and disposing of organic waste. Disinfect the mushroom house with Dicofol (0.1%). Spray the compost with Diazinon @ 1.5-2 ml / 10L water.

Nematodes: - Fungi: Sources of infection:

Compost ingredients such as wheat straw, chicken manure, horse manure, sawdust, wooden bowls, shelves and other containers, etc .; may be the primary source of infection. Distribution: Once these nematodes enter the mushroom house, they spread further through the air, with faulty water spray, working hands, equipment, mushroom flies, mites, and so on. There are two types of fungal nematodes: mycophagus or myceliophagic nematodes

Mycelial nematodes (Aphelenchoides composticola, Aphelenchoides agarici, A. neocomposticola, Ditylenchus myelophagous) These nematodes feed directly on the fungal mycelium and fruit bodies. They are fitted with special types of mouthpieces, namely H. Style or needle used by these parasites to pierce the vagina, inject digestive juices and suck the contents of the cell.

Soapworms (Rhabditis spp., Panagrolaimus spp. Diplogaster spp.): Instead of their style they contain a tubular mouthpiece through which the nutrient particles of the substrate are sucked in. With their tubular mouths, they are not capable of causing direct damage to the fungus. They deteriorate the structure and quality of compost in cultivated areas and cause bad odors, reduced yields, shorter harvest time and quantitative loss of sporophore, etc.

Symptoms of nematode infection: The surface of the compost sinks to the mycelium and grows moderately and becomes tired. The white mycelium disappears

Leave the infected mushroom compost only with coarse fibers with black compost mass. Due to the accumulation of a large number of bacteria, the compost will be moist and will have an unpleasant odor. The taps turn brown, wet and stumble. The fruit bodies

appear as patches on the beds. The yield is drastically reduced due to the rinsing pattern and the harvest time.

Control Methods: Full Hygiene Proper pasteurisation of compost and coverings Soak the mushroom house and buildings with some disinfectant. Using fresh polyethylene bags and sterilizing empty trays or carts with formalin or other disinfectant, rugoso-annulata, etc.

Fungi that intercept nematodes, such as Arthrobotrys oligospora, A. superba and A. robusta. Different types of pleurotus can be used as biocontrol agents against fungal nematodes. Neem, Castor, Groundnu, Karanj, etc. Mixing plant extracts in compost during spawning or cultivation.

COMMON PROBLEMS WITH MUSHROOM

1. Not enough moisture

The fungus, the underground vegetative growth of the fungus, requires a moist environment for the fungus to bloom and produce. The fungus consists mainly of water. So if you let the mycelium dry or if the humidity is too low, nothing will happen.

Make sure the mushroom carrier is moist enough. Do you see the picture on the right? In summer I made this outdoor mushroom project out of cardboard, straw and mycelium. Then I stroked a good job, went on vacation and forgot for a while.

As you can imagine, in the hot July temperatures, everything dried up. I had hard work and guilt for baking my bad mycelium.

Solution: Pay attention to moisture and humidity! When growing mushrooms outdoors, make sure that your bags or bed are slightly damp. Make sure the fog or water dries out when you see and feel things.

If the interior has been cultivated under sterile conditions, you should carefully check the moisture and moisture content. A cheap moisture meter will help you with this.

2. Too much moisture

The opposite of the above problem is happening. Too much moisture can lead to wet surfaces, mold and stagnant water.

Stagnant water promotes bacterial growth and mold, two things that compete with mycelium. Although we want to keep our growing medium humidity and soak it for a day at first, it is only problematic to leave it in stagnant water.

Solution: It's all about drainage. If using a mushroom growing kit, do not leave it in the water after spraying. If you use a bag or container or other method for indoor use, do not spill into water and drill holes in the floor to allow water to leak out.

Remember this when trying to grow your own mushroom outdoors. The bed should be properly drained and should not be placed where it is in the water and carries mold

3. Not Sterile Enough

The microbial world is a constant battle of good versus evil. Your mycelium needs to take over and remain in control of your substrate, or it will lose out to mold and other micro-competitors.

Failing to take this into account will lead to bugs, mold spores, and other unhealthy things taking over your project. Even if it does produce mushrooms, you probably won't want to eat them.

Use an autoclave to grow your own mushrooms indoors Solution: This is often easier to do when trying to grow mushrooms in outdoor beds. Keep cleanliness in mind by maintaining a good working environment.

Follow obvious rules like washing your hands and not working next to the litter box.

Depending on the type of project, you may want to prepare your substrate first to discourage micro-competitors. Pasteurization of straw is one of these methods.

For some indoor projects like growing from spores you'll want to maintain strict sterility in order to avoid contamination. Getting equipment such as a flow hood, autoclave, or pressure cooker (right) is often necessary. Do a lot of reading before you do something like this. It's not for beginners!

4. Not Enough Air Exchange

Mushrooms don't need as much fresh air as we do, but they still need it. Without any air exchange carbon dioxide levels build up and your mushrooms will emerge as stunted, spindly things that are all stalks and no caps. Very disappointing.

Solution: Make sure your project has a flow of fresh air. Don't place things in areas with no air exchange.

If you're growing in a sealed environment, you may want to open it a few times a day for fresh air. Just be aware that when you introduce fresh air you also introduce the possibility of contaminants and lower humidity levels. It's a delicate dance!

5. The Wrong Environment

The key to learning how to grow mushrooms is to create an environment that's conducive to the species that you're cultivating. Make them feel at home!

That means don't try to grow a warm-temperature mushroom in cold weather. Don't try to cultivate a wood loving species on straw. Make sure your mushroom substrate is nutrient rich. Basically, give the mycelia what it needs to thrive.

Solution: Research. Know what kind of mushroom you're trying to grow and what they need before you begin. You don't have to read someone's PhD thesis, but a little knowledge goes a long way

6. Bad Spawn

Mushroom spawn that's old or has traveled a great distance may not be as vigorous and may fail to thrive and produce. It's no great secret that you should have the healthiest spawn possible to increase your chances to successfully grow mushrooms.

Solution: First off, only buy spawn from a reputable company. If you purchase from someplace sketchy with bad business practices you'll get an inferior product. Ideally buy from somewhere close to you, so your spawn doesn't have to go very far.

After that the best advice is to use it or lose it! Don't let spawn sit around forever, as it will weaken, create wastes, and possibly contaminate. Keeping it in the refrigerator will extend its life, but it becomes less viable with every passing week.

7. Lack of Research/Understanding of the Mushroom Life Cycle

Understand the mushroom life cycle and help your mycelium grow! You don't have to be a professional mycologist to understand some basic principles of the mushroom life cycle. Knowing how this organism works greatly decreases the chances of your mushrooms not growing. You'll be better equipped if you understand what mycelium is, how it feeds itself, and what it needs to survive.

Solution: Again, research. You don't have to know everything, but some research in the beginning is important.

8. Lack of Patience

Mycelium takes time to grow into a substrate and grow mushrooms. In the case of some mushrooms, like morels, it may even take years!

This is not an activity for the impatient, something I struggle with as a fairly impatient person myself. Yet fear not, careful watching and waiting is greatly rewarded in this hobby.

PROCESSING METHOD OF MUSHROOM FROM DRYING TO FREEZING

Drying

One of the simplest and most reliable ways of preserving mushrooms is by drying them. Mushrooms should be sliced first, to quicken the process. If you don't have a food dehydrator, you can put them on an aluminum foil tray over an electric lamp (or something else with a old-style light bulb you can just keep on) or the pilot light on the stove. My own stove has two hot spots from pilot lights that work very nicely; but my lights have all been switched over to fluorescent, so that doesn't work for me anymore.

Mushrooms should be dried until they have the texture of a crisp potato chip. If they are still flexible, they will continue to rehydrate and... bad things will happen. This also means that they will need to be stored in air-tight containers. But if these steps are followed correctly, they will keep for years. Some museum herbaria still have dried mushroom specimens from the 1700s, though I'm not sure I'd want to eat them.

Some mushrooms get tough or otherwise get a bad texture when dried. Chanterelles get quite leathery, for example. These mushrooms can be dried and then powdered and added to dishes as a flavoring – or of course they can be preserved in other ways. Dried slices of giant puffball can be used to build a pizza on, or they can be crumbled into soup for a nice fungal flavor and thickening. Drying is the traditional way to preserve morels and boletes, and it strengthens their flavor.

Freezing

Freezing is the other preservation method that comes naturally to most of us. The main point here is that all mushrooms should be sautéed (or parboiled, or briefly cooked in some other way) before being frozen. Otherwise, unless you really, really flash freeze (and unfreeze) them, they usually liquefy upon thawing – a most unpleasant experience. Oyster mushrooms are absolutely the worst for this, which is really strange since in nature oyster mushrooms can freeze solid on the tree, and you thaw them out and they're fine. But if you freeze them in your freezer without sautéing first, you end up with mush when you thaw them.

This is also good for some mushrooms, like hen of the woods, that have a reputation for needing a long time to cook. In my experience, most of those mushrooms need some cooking at a higher heat than they often get in order to soften up. For instance, the hen of the woods one sees with a recommendation for soups or casseroles. But if you sauté it for a few minutes in a mixture of butter and oil, you can raise it to a temperature beyond what it usually reaches with those methods, and this softens it right up and you can just put it into your soup then and have it good to go. Or, likewise, you can freeze it then and have it cooked and ready to use as it comes out of the freezer. Also, you don't have to sauté your mushrooms on their own – you can add some onions and garlic to the pan and make a mixture called duxelles that also freezes very well and is a great ready-to-use ingredient.

Other methods

More elaborate preservation methods are of course possible: pickling, confiting, smoking are all good ways to make tasty preserved mushrooms. They also are more elaborate and take more work, and change the flavor of the mushroom more. With drying and freezing, you're preserving the mushroom pretty much as much as possible with its original flavor. The two methods can be combined, to a certain extent: if you want to make absolutely sure of your dried mushrooms, you can put them in a deep freeze for a while each year (or microwave them) to kill any bug eggs or larvae that might be on them. Since your mushrooms are potato-chip crisp and completely dehydrated, they will not be affected by the microwaving (or freezing), right?

In any case, I hope this helps you enjoy your mushrooms.

www.ingramcontent.com/pod-product-compliance
Lightning Source LLC
Chambersburg PA
CBHW030622220526
45463CB00004B/1379